Windows Phone 7.5 Application Development with F#

Develop amazing applications for Windows Phone using F#

Lohith G.N.

PUBLISHING

BIRMINGHAM - MUMBAI

Windows Phone 7.5 Application Development with F#

First published: April 2013

Production Reference: 1030413

Published by Packt Publishing Ltd.
Livery Place
35 Livery Street
Birmingham B3 2PB, UK.

ISBN 978-1-84968-784-3

www.packtpub.com

Cover Image by Siddharth Ravishankar (sidd.ravishankar@gmail.com)

Credits

Author

Lohith G.N.

Reviewers

Senthil Kumar

Vivek Thangaswamy

Acquisition Editor

Kevin Colaco

Commissioning Editor

Priyanka Shah

Technical Editors

Worrell Lewis

Lubna Shaikh

Copy Editors

Brandt D'Mello

Alfida Paiva

Laxmi Subramanian

Ruta Waghmare

Project Coordinator

Esha Thakker

Proofreader

Elinor Perry-Smith

Indexer

Hemangini Bari

Graphics

Aditi Gajjar

Production Coordinators

Manu Joseph

Nitesh Thakur

Cover Work

Manu Joseph

About the Author

Lohith G. N. hails from Mysore, India and currently resides in Bangalore, India. He has over 12 years of experience in software development. He presently works as a Developer Evangelist for Telerik in India and takes care of Evangelism for the South Indian region. He comes from a Production Engineering background and ended up in software development thanks to the FORTRAN language that he learned during his graduation days. Being well versed with the .NET platform, Lohith has experience building web applications, Windows applications, and Service Oriented Architecture. He has spent close to a decade mostly in the services-based industry and is well versed with the agile method of software development.

Lohith is also a two time Microsoft Most Valuable Professional (MVP) in the area of ASP.NET/IIS. He was given this prestigious award from Microsoft in 2011 and 2012. He often writes on ASP.NET/ODATA and maintains his own blog at `http://kashyapas.com`. He can be reached on Twitter and his Twitter handle is `@kashyapa`. To know more about Lohith you can check out `http://about.me/kashyapa`. Lohith is also one of the User Group leads for Bangalore DotNet User Group — one of the most active User Groups in India. He is a regular speaker at the local user groups.

This is the first ever book that I have written and I take this opportunity to thank my parents. I would also like to thank my lovely wife Rashmi and my lovely son Adithya for having put up with me while writing this book. I have promised them a nice vacation as soon as I am done with the book.

About the Reviewers

Senthil Kumar is a Software Engineer and a passionate blogger. He works mainly on Windows or client development technologies and has good working experience in C#/.NET, Delphi, Win forms, Windows Phone, Windows 8, and SQL Server.

He completed his Master of Computer Applications from Christ College (Autonomous), Bangalore in the year 2009 and is a MCA rank holder (gold medalist). He has worked as a technical reviewer for *Windows Identity Foundation Cookbook, Sandeep Chandra, Packt Publishing*.

You can connect with him on Twitter at `http://twitter.com/isenthil`, on Facebook at `http://www.facebook.com/kumarbsenthil`, and his blog at `www.ginktage.com`.

Vivek Thangaswamy is highly committed to technology support and service to the global community and workplace. Looking at his community support activity Microsoft awarded him the Most Valuable Professional (MVP) award for three consecutive years—2007 for ASP.NET, 2008 and 2009 for SharePoint. He has also been awarded the Professional Excellence and Innovation Award for the year 2011 from `www.npa.org`. He has been awarded with bronze, silver, and gold medals by `dotnetspider.com` for his contribution to the community. `www.experts-exchange.com` has awarded him Master status in XML, ASP.NET, and SharePoint.

Apart from this recognition, Vivek has contributed to the MSDN forums and www. codeproject.com. He is the administrator for all technology-related discussions at www.redpipit.com. He is the creator for two projects in www.codeplex.com—an open source community for Microsoft Technologies. He has authored *Exploring SharePoint Foundation 2010, Darkcrab Press* and *VSTO 3.0 for Office 2007 Programming, Packt Publishing,* co-authored the book *System Analysis and Design, LAP Lambert Academic Publishing,* and has been the technical reviewer for three books *Microsoft Office Live Small Business: Beginner's Guide, Packt Publishing, Refactoring with Microsoft Visual Studio 2010, Packt Publishing,* and *BlackBerry Enterprise Server 5 Implementation Guide, Packt Publishing.*

I would like to dedicate this book to my family and friends; they are the confidence and the strength in my life.

www.PacktPub.com

Support files, eBooks, discount offers and more

You might want to visit www.PacktPub.com for support files and downloads related to your book.

Did you know that Packt offers eBook versions of every book published, with PDF and ePub files available? You can upgrade to the eBook version at www.PacktPub.com and as a print book customer, you are entitled to a discount on the eBook copy. Get in touch with us at service@packtpub.com for more details.

At www.PacktPub.com, you can also read a collection of free technical articles, sign up for a range of free newsletters and receive exclusive discounts and offers on Packt books and eBooks.

http://PacktLib.PacktPub.com

Do you need instant solutions to your IT questions? PacktLib is Packt's online digital book library. Here, you can access, read and search across Packt's entire library of books.

Why Subscribe?

- Fully searchable across every book published by Packt
- Copy and paste, print and bookmark content
- On demand and accessible via web browser

Free Access for Packt account holders

If you have an account with Packt at www.PacktPub.com, you can use this to access PacktLib today and view nine entirely free books. Simply use your login credentials for immediate access.

Instant Updates on New Packt Books

Get notified! Find out when new books are published by following @PacktEnterprise on Twitter, or the *Packt Enterprise* Facebook page.

Table of Contents

Preface

Windows Phone 7.5 Application Development with F# is a book for anyone who is familiar with F# and wants to try a hand at Windows Phone application development. This book will cover the basics of application building on the Windows Phone platform but using F# as the language. We will cover everything, from basic requirements to programming on the platform, to project templates, and to developing screens. This book will act as a ready reckoner for folks who want to quickly look at the concepts of Windows Phone application programming.

What this book covers

Chapter 1, Setting up Windows Phone Development with F#, is all about setting the stage for Windows Phone development with F#. Here we will take a look at the Windows Phone platform, the F# language, and the prerequisites required to start developing applications.

Chapter 2, F# Windows Phone Project Overview, is all about becoming familiar with the different project templates available for developing Windows Phone applications. We will decipher each project template and understand what each project type contains and how to work with each type. We will also look at some of the item templates required for app development.

Chapter 3, Working with Windows Phone Controls, helps us understand the controls provided by the platform. We will look at more than 10 controls provided by this platform. We will take one control and walk you through how to work with that control. By the end of this chapter you will be familiar with the "toys" you can use to play on this platform.

Chapter 4, *Windows Phone Screen Orientations*, introduces you to a concept called orientation and shows you how to deal with it in your applications. Since a phone is a handheld device, the user has all freedom to turn the phone upside down or rotate it left or right. This changes the orientation of your application, and as the developer, it is your responsibility to react to this. This chapter will help you understand the different orientations and how to code for handling orientation changes.

Chapter 5, *Windows Phone Gesture Events*, teaches you how to read the gestures performed by the user in your application. We will take a look at what gestures are and what gesture events are supported by the platform. We will also look at how to handle gesture events in your applications.

Chapter 6, *Windows Phone Navigation*, is all about understanding how to allow users to move from one screen to another screen in your application. We will look at the Windows Phone navigation model and different techniques to enable navigation.

Chapter 7, *Windows Phone and Data Access*, helps you understand how to store and access data on the Windows Phone platform since one of the fundamental aspects of any application is data. We will try to understand the different data source options available on the platform.

Chapter 8, *Launchers and Choosers*, introduces a whole set of built-in applications, also known as Launchers and Choosers. Launchers and Choosers help us make use of the built-in apps or call the built-in apps right from our own apps.

Chapter 9, *Windows Phone Sensors*, introduces you to multiple sensors supported by Windows Phone that allow apps to determine orientation and motion of the device. With sensors, it is possible to develop apps that make the physical device itself an input. Typical uses of these sensors are for games, location-aware apps, and so on. The Windows Phone platform provides APIs to retrieve data from the individual sensors.

What you need for this book

In order to work through this book and to learn Windows Phone 7.5 Application Development with F#, you will need to have the following software:

- Visual Studio 2010
- Windows Phone Software Development Kit 7.1
- Windows Phone Project and Item Template for F#

Who this book is for

If you know F# and are interested in developing for the Windows Phone 7.5 platform, this book is for you. It gives you a jump-start to developing Windows Phone 7.5 apps using F#.

Conventions

In this book, you will find a number of styles of text that distinguish between different kinds of information. Here are some examples of these styles, and an explanation of their meaning.

Code words in text are shown as follows: "This contains the launch point, which is App.XAML."

A block of code is set as follows:

```
public AppHost()
    {
        // Standard Silverlight initialization
        InitializeComponent();
        app = new WindowsPhoneApp.App(this);
    }
```

New terms and **important words** are shown in bold. Words that you see on the screen, in menus or dialog boxes for example, appear in the text like this: "Click on **OK** once you are done with naming the application."

 Warnings or important notes appear in a box like this.

 Tips and tricks appear like this.

Reader feedback

Feedback from our readers is always welcome. Let us know what you think about this book—what you liked or may have disliked. Reader feedback is important for us to develop titles that you really get the most out of.

To send us general feedback, simply send an e-mail to `feedback@packtpub.com`, and mention the book title via the subject of your message.

If there is a topic that you have expertise in and you are interested in either writing or contributing to a book, see our author guide on `www.packtpub.com/authors`.

Customer support

Now that you are the proud owner of a Packt book, we have a number of things to help you to get the most from your purchase.

Downloading the example code

You can download the example code files for all Packt books you have purchased from your account at `http://www.packtpub.com`. If you purchased this book elsewhere, you can visit `http://www.packtpub.com/support` and register to have the files e-mailed directly to you.

Errata

Although we have taken every care to ensure the accuracy of our content, mistakes do happen. If you find a mistake in one of our books—maybe a mistake in the text or the code—we would be grateful if you would report this to us. By doing so, you can save other readers from frustration and help us improve subsequent versions of this book. If you find any errata, please report them by visiting `http://www.packtpub.com/submit-errata`, selecting your book, clicking on the **erratasubmissionform** link, and entering the details of your errata. Once your errata are verified, your submission will be accepted and the errata will be uploaded on our website, or added to any list of existing errata, under the Errata section of that title. Any existing errata can be viewed by selecting your title from `http://www.packtpub.com/support`.

Piracy

Piracy of copyright material on the Internet is an ongoing problem across all media. At Packt, we take the protection of our copyright and licenses very seriously. If you come across any illegal copies of our works, in any form, on the Internet, please provide us with the location address or website name immediately so that we can pursue a remedy.

Please contact us at copyright@packtpub.com with a link to the suspected pirated material.

We appreciate your help in protecting our authors, and our ability to bring you valuable content.

Questions

You can contact us at questions@packtpub.com if you are having a problem with any aspect of the book, and we will do our best to address it.

1
Setting up Windows Phone Development with F#

In this chapter, we will try to understand the three important aspects that make up Windows Phone Development using F# (pronounced as F sharp), namely:

- What is Windows Phone?
- What is F#?
- Prerequisites for development

We will go over these aspects, one by one in the coming sections.

What is Windows Phone?

Windows Phone is the new mobile operating system from Microsoft Corporation and was launched in October 2010. After the initial release, there were a series of updates to Windows Phone with the recent one being Windows Phone 7.5 (Mango).

Windows Phone has the tagline "Put people first", and is mainly aimed at consumers or end users. Windows Phone is the successor to a previous version of mobile operating system from Microsoft known as Windows Mobile. **Windows Mobile** was an operating system designed around the Windows CE (Compact Edition) kernel. Windows Phone, being a new platform written from scratch, is not compatible with the earlier versions of Windows Mobile, that is, it does not support backward compatibility. So applications written for Windows Mobile won't run on Windows Phone. Windows Phone and Windows CE are just two different mobile platforms available at present from Microsoft.

Windows Phone has a fresh and new user interface called **Modern UI**, a typography-based design language that is inspired by the transport system.

Windows Phone – a standardized platform

The biggest problem that application developers for mobile platforms faced was the varied range of development environments they had to adapt to. The mobile development environment was completely different from those compared to either a desktop application development environment or a web application development environment. Though some development environments like Microsoft Platform, which includes developing using the popular **Integrated Development Environment (IDE)** Visual Studio and languages like Visual C++ or Visual C# were, to some extent, similar.

But, one had to face the challenges of handling different form factors, device capabilities, hardware differences, and other incompatibilities. With Windows Phone, Microsoft has made sure that it provides a common design and a common set of capabilities for devices from many different manufactures. So be it any device from any manufacturer, as a developer we only have one set of design and capabilities to tackle. This makes it easier for the developers to concentrate on their application and not worry about any other nuances.

Microsoft has set minimum requirements for the hardware on which the Windows Phone runs. The hardware requirements are as follows:

- **Capacitive touch**: Four or more contact points
- **Sensors**: GPS, accelerometer, compass, light, proximity
- **Camera**: 5 MP or more, Flash, dedicated camera button
- **Memory**: 256 MB, 8 GB flash storage or more
- **GPU**: DirectX 9 acceleration
- **Processor**: ARMv7 Cortex/Scorpion or better
- **Screen sizes**: 480 x 800 WVGA, 480 x 320 HVGA
- **Keyboard**: Optional
- **Hardware buttons**: Must be fixed on the face

The following image from `http://msdn.microsoft.com/en-us/library/gg490768.aspx` shows the various features a Windows Phone has to offer for both developers as well as consumers:

Design
Contemporary modern style
Intuitive gesture-driven UI
"Metro" power-saving theme
Animated page transitions
Highly interactive feedback
Horizontal and vertical display
Optional hardware keyboard

Minimum Specification
800 x 480 WVGA screen
Graphics processor unit
DirectX9 acceleration
Dedicated media codecs
256 MB or more memory
8 GB or more flash memory
ARM V7 Cortex/Scorpion processor

Input
On-screen keyboard
Camera with flash
Microphone

Media Features
Integrated media library
Video playback
Musica and playlists
Photo library
Animation playback
FM radio

Messaging
Email integration
SMS text messaging
Windows Live

Intuitive Controls
Panorama and pivot views

Communication
Phone
Contacts list
Social media sites

Push Notifications
On-screen message
Program tile updates
Raw notifications

Applications
Games
Windows Marketplace
Custom applications

Application Bar
Application-specific buttons
Text captions available

Connectivity
3G and GPRS services
Wireless (Wi-Fi)
Bluetooth
Micro SD connector
Micro USB connector

Back button

Start screen

Intelligent search

Sensors
Accelerometer
GPS and Wireless location
Multi-touch screen
Proximity sensor
Compass
Light sensor

Windows Phone features

Development option for Windows Phone

Windows Phone Application Platform is built on the existing Microsoft tools and technologies, such as Visual Studio, Expression Blend, Silverlight, and XNA Framework. The learning curve for developing Windows Phone Application is minimal for anyone who is familiar with the tools and technologies on which the platform is built. Windows Phone Application Platform provides two main frameworks for development. They are:

- **Silverlight Framework** – used for event-driven, XAML-based applications. This framework allows developing markup-based and rich, media-based applications.

- **XNA Framework** – used for loop-based games. Allows developing immersive and fun gaming- and entertainment-based applications.

Windows Phone Application Platform Architecture

The platform itself is made up of four main components. The following figure from `http://i.msdn.microsoft.com/dynimg/IC513005.jpg` shows the components of Windows Phone Application Platform:

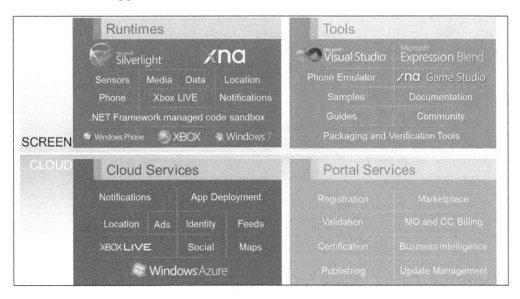

Windows Phone Application Platform

In this figure we have:

- **Runtimes**: Silverlight, XNA Framework, and phone-specific features provide the environment to build graphically-rich applications
- **Tools**: Visual Studio, Expression Blend, and related tools provide developer experience to create, debug, and deploy applications
- **Cloud Services**: Azure, XBOX Live, notification services, and location services provide data sharing across cloud and a seamless experience across the devices a consumer will use
- **Portal Services**: Windows Phone Marketplace allows developers to register, certify, and market their applications

The components of interest for any developer are runtime and tools. Runtime because that's the base on which applications are developed. Tools play another major part in the development experience. Visual Studio and Expression Blend try to enhance the development experience by providing features that makes a developer's job easy while developing. Visual Studio in particular is a well-known IDE, which lets you create, debug, and deploy an application without having to go out of the IDE.

All phases of the development can be achieved staying within the IDE and this is the biggest experience one gets when on this platform. Expression Blend makes visual designing very easy as it allows the drag-and-drop capability on the design surface. When designing in Blend, you just set a bunch of properties and the code is automatically written by the Blend for you.

What is F#?

F# is a .NET programming language. F# was initially started as a research project at Microsoft Research Lab by Don Syme. Later, it became a mainstream .NET programming language and is distributed as a fully supported language in the .NET Framework as part of Visual Studio.

According to Microsoft Research, F# is:

> *A succinct, expressive, and efficient functional and object-oriented language for .NET that helps you write simple code to solve complex problems.*

F# is a strongly typed language; it uses the type inference. Since it uses the type inference, programmers need not declare the data types explicitly. The compiler will deduce the data type during compilation. F# will also allow explicit declaration of data types.

Prerequisites for development

To start developing for Windows Phone using F# as a language, you will need some prerequisites to be installed on your development system. The prerequisites are as follows:

- Visual Studio 2010
- Windows Phone Software Development Kit 7.1
- Windows Phone Project and Item Templates for F#

So let's take a look at these one by one.

Visual Studio 2010

When you install Visual Studio 2010 (Professional or Ultimate) and choose the default options during installation, the installer will, by default, install Visual C#, Visual C++, Visual Basic, and Visual F#. This is by far the easiest way of starting to develop with F#.

Windows Phone Software Development Kit 7.1

The **Software Development Kit (SDK)** provides us with the tools needed to develop applications and games for the Windows Phone platform. The SDK can be downloaded from `http://gnl.me/WPSDK71`. The SDK installs the following components on your development system:

- Microsoft Visual Studio 2010 Express for Windows Phone
- Windows Phone Emulator
- Windows Phone SDK 7.1 Assemblies
- Silverlight 4 SDK and DRT
- Windows Phone SDK 7.1 Extensions for XNA Game Studio 4.0
- Microsoft Expression Blend SDK for Windows Phone 7
- Microsoft Expression Blend SDK for Windows Phone OS 7.1
- WCF Data Services Client for Window Phone
- Microsoft Advertising SDK for Windows Phone

Windows Phone project and Item Templates for F#

The easiest way to get up and running is to utilize one of the project templates available through the Visual Studio Gallery. These templates have been created by the F# community and they provide a great way to kick-start your project. Daniel Mohl, an F# **Most Valuable Professional** (**MVP**) has written a couple of Visual Studio templates that will help us to quickly get up and running with Windows Phone development using F#. You can download any of the following available templates based on your needs. The templates and URL from where you can download them is as follows:

- F# and C# Windows Phone App (Silverlight) Template: `http://gnl.me/FSharpWPAppTemplate`

- F# and C# Windows Phone List App (Silverlight) Template: `http://gnl.me/FSharpWPListAppTemplate`

- F# and C# Windows Phone Panorama App Template: `http://gnl.me/FSharpWPPanoramaAppTemplate`

- F# XAML Template: `http://gnl.me/FSharpXAMLTemplate`

Summary

In this chapter you learned about Windows Phone as a new mobile platform. We looked at how Windows Phone Application Platform offers a standardized platform for developers. We also looked at several features that Windows Phone provides.

Then we looked at a new functional programming language in .NET framework called F#. F#, which started in Microsoft Research Lab is now a mainstream .NET programming language.

After understanding Windows Phone as a platform and F# as a language, we then looked at the prerequisites that are required for developing applications for Windows Phone using F# as the language.

In the next chapter we will take a look at the different project templates we downloaded. We will go in depth into each project and understand the various components of each project.

2
F# Windows Phone
Project Overview

The previous chapter gave us a peek into Windows Phone as a Mobile Platform, F# as a language, and the prerequisites required for developing Windows Phone applications using F#. In this chapter, we will learn more about the three F# Windows Phone Application Project templates, which are required for developing the apps.

Windows Phone Project Templates for F#

In the previous chapter, we read about three project templates necessary for Windows Phone Application development with F#. They are:

- F# and C# Windows Phone Application (Silverlight) Project Template
- F# and C# Windows Phone List Application (Silverlight) Project Template
- F# and C# Windows Phone Panorama Application (Silverlight) Project Template

Once you have downloaded and installed the prerequisite project templates, go to **File | New Project in Visual Studio IDE**. You should see a **Project** dialog box, which will show the Windows Phone Project Template for F#, as follows:

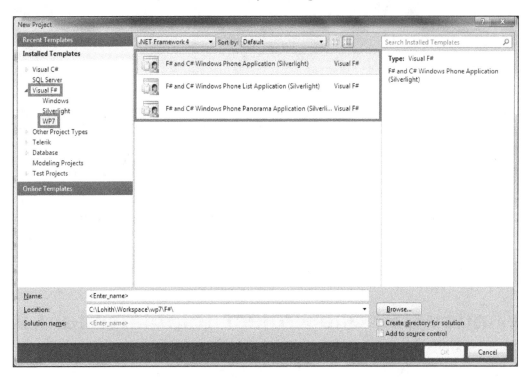

In the upcoming section, we will go over each project type, and understand how it works and what it contains in terms of project items. We will create projects of all the above listed types and take a closer look at each one of them.

F# and C# Windows Phone Application (Silverlight) Project Template

Open an instance of Visual Studio IDE. From the standard menu, select **File | New Project**. In the **New Project** dialog window, select **Other Languages | Visual F# | WP7** under the **Installed Templates** section. In the middle section, select **F# and C# Windows Phone Application (Silverlight)**. Give a name for the app and select a location for the project to be saved. Click on **OK** once you are done with naming the application. Visual Studio will go ahead and create a basic Windows Phone Application project with F# as the language. Once the solution has been created successfully, it should look similar to the following screenshot:

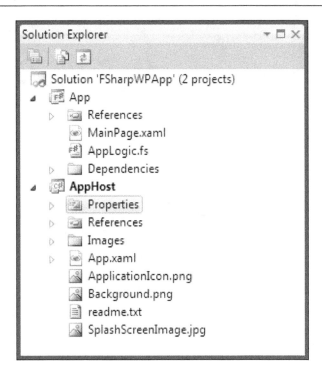

Now, let's try to understand what's in the solution. As you can see from the screenshot, we have two projects in the solution:

- **AppHost**
- **App**

The AppHost project

AppHost is the main project and also the startup project in the solution. This project is of the type Silverlight Application for Windows Phone and uses C# (C Sharp) as its language. As the name goes, this host project references another project in the solution named App. This contains the launch point, which is App.XAML. This project also contains the bare minimum requirements of a Windows Phone Application, that is, splash screen image, application icon, background for tiles, and the Manifest file.

If you right-click on this application and select **Properties**, you can visualize the project properties in a GUI, as follows:

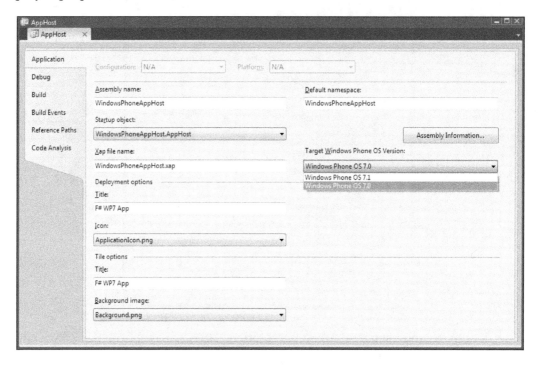

 At the time of writing this book, there are two versions of the Windows Phone SDK that an application can target to, namely, Windows Phone OS 7.0 and Windows Phone OS 7.1. The "F# and C# Windows Phone Application (Silverlight)" project template is, by default, setting the project target OS to 7.0. You will need to select the OS as 7.1 and save the project in order to target the App to the latest Windows Phone OS.

By default, the project template names the namespace, assembly, and XAP filename as WindowsPhoneAppHost. If you don't like the default naming, you can rename them according to your policies, and do a rebuild on the project.

Let's open the App.xaml.cs file (which is also known as code behind of App.xaml) and understand what happens at the start of the project. Here is what the code looks like in App.xaml.cs:

```
using System;
using System.Linq;
```

```
using System.Windows;
using Microsoft.Phone.Controls;

namespace WindowsPhoneAppHost
{
  public partial class AppHost : Application
  {
    // Easy access to the root frame
    public PhoneApplicationFrame RootFrame { get; private set; }
    WindowsPhoneApp.App app;
    // Constructor
    public AppHost()
    {
      // Standard Silverlight initialization
      InitializeComponent();
      app = new WindowsPhoneApp.App(this);
    }
  }
}
```

AppHost inherits the System.Windows.Application class. It encapsulates a Silverlight application because of this inheritance. It defines two properties—one for RootFrame and another for App. Every Windows Phone Application contains a root frame and all the visual content is painted on this frame, so the application gives us a handle to that frame. WindowsPhoneApp.App is a class in another project in the solution called App. We will look at this project in the next section. When AppHost starts within the constructor, the standard Silverlight application initialization happens to call and initialize the App class.

Another interesting aspect of this project is the manifest file. This file is known as WMAppManifest.xml and can be found in the Properties folder. Every Windows Phone App should contain a manifest file. As the name suggests, it contains certain metadata so that the Windows Phone OS can know more about our application. The code of our manifest file is as follows:

```
<?xml version="1.0" encoding="utf-8"?>
<Deployment xmlns=
  "http://schemas.microsoft.com/windowsphone/2009/deployment"
```

```
      AppPlatformVersion="7.0">
    <App xmlns="" ProductID="{0056bd0c-d7a2-4f93-9f1f-8ee5bbef8c76}"
      Title="F# WP7 App" RuntimeType="Silverlight"
      Version="1.0.0.0" Genre="apps.normal" Author="WindowsPhoneApp
      author" Description="Sample description"
      Publisher="WindowsPhoneApp">
    <IconPath IsRelative="true" IsResource="false">
      ApplicationIcon.png
    </IconPath>
    <Capabilities>
      <Capability Name="ID_CAP_GAMERSERVICES"/>
      <Capability Name="ID_CAP_IDENTITY_DEVICE"/>
      <Capability Name="ID_CAP_IDENTITY_USER"/>
      <Capability Name="ID_CAP_LOCATION"/>
      <Capability Name="ID_CAP_MEDIALIB"/>
      <Capability Name="ID_CAP_MICROPHONE"/>
      <Capability Name="ID_CAP_NETWORKING"/>
      <Capability Name="ID_CAP_PHONEDIALER"/>
      <Capability Name="ID_CAP_PUSH_NOTIFICATION"/>
      <Capability Name="ID_CAP_SENSORS"/>
      <Capability Name="ID_CAP_WEBBROWSERCOMPONENT"/>
    </Capabilities>
    <Tasks>
      <DefaultTask Name="_default" />
      <!--NavigationPage="/WindowsPhoneApp;component
        /MainPage.xaml" -->
    </Tasks>
    <Tokens>
      <PrimaryToken TokenID="WindowsPhoneAppHostToken"
        TaskName="_default">
        <TemplateType5>
          <BackgroundImageURI IsRelative="true"
            IsResource="false">
            Background.png
          </BackgroundImageURI>
          <Count>0</Count>
          <Title>F# WP7 App</Title>
        </TemplateType5>
      </PrimaryToken>
    </Tokens>
  </App>
</Deployment>
```

The App project

App is an F# project. It is basically like an F# library. The output of this project, which is a DLL, is referenced in the AppHost project.

This project contains two main components. One is `MainPage.xaml` — the screen or UI or visual for the sample project. The contents of XAML are nothing but screen elements positioned at specific points on the screen. This does not contain any code behind file, that is, `MainPage.xaml.cs`.

The second component of the project is `AppLogic.fs`. Pretty similar to a C# class file, this is an F# class file with an `.fs` extension. This file contains all the logic required for our project. So let's see the important logic hidden inside this class one by one:

- **Module utilities**: This module contains a helper utility to reference visual elements in the `.xaml` file, by name in the F# class.

- **Type MainPage**: This is the class that handles pretty much everything off of `MainPage.xaml`. It inherits `PhoneApplicationPage`. It is responsible for loading the XAML and bringing up the UI. It also handles other things, such as button click and app bar button clicks.

- **Type App**: This is the class that handles the application aspects of our project. One instance of this is created in the `AppHost` instantiation and it handles some of the housekeeping on the `AppHost` itself. These include navigating to `MainPage`, and handling the application lifetime events, such as launching, closing, activated, and deactivated.

Output

If you build the solution without making any changes at this point, the build will succeed. So, you can run the app and watch a Windows Phone emulator launching and running the app inside the emulator. Here is a screenshot of the app we just built:

The button1 and button2 click events are wired to an event handler inside type MainPage. When you click on these buttons, a message will be shown on the main page.

So, we saw a basic Windows Phone Application project template for F#. We understood the different project composition of the solution, different components inside a project, and finally the output. In the following sections, we will cover the remaining two types of project templates, namely, the List app and Panorama apps.

F# and C# Windows Phone List Application (Silverlight) Project Template

The F# and C# Windows Phone List Application (Silverlight) Project Template is another project template available for Windows Phone Application development with F#. Create a new project and select the **F# and C# Windows Phone List Application (Silverlight)** project template. Provide a name for your **Project**, select the location, and click on **OK**. The solution will be created, and will be similar to what we saw in the previous template.

This solution also has the two familiar projects, namely AppHost and App. As described in the previous section, the content of AppHost in this solution is also the same. App contains the same AppLogic.fs F# class file. As the name of the template goes, this is an example of the List application. This project contains two XAMLs, namely, MainPage and DetailsPage. MainPage will show a list, and clicking on each of its item will display its details. The AppLogic class now contains MainPage and DetailsPage to deal with the business logic for these two XAMLs. The following is an output of this sample project:

F# and C# Windows Phone Panorama Application (Silverlight) Project Template

The third of the available project templates for F# Windows Phone Application development is "F# and C# Windows Phone Panorama Application (Silverlight)". As the names goes, this sample project shows how to use a Panorama control in a Windows Phone App. **Panorama** is a control that allows multiple pages to be hosted, but only one page is shown at any time. A user can flick to the left or right to see the other pages.

As shown in the previous two sections, this project also contains the same project composition, namely `AppHost` and `App`. One additional aspect of this project is the use of a third-party framework known as **Caliburn Micro**, which is a small yet powerful framework, designed for building applications across all XAML platforms. Caliburn Micro supports a coding pattern known as **Model View ViewModel (MVVM)**. Caliburn Micro uses convention over configuration. For more information on how Caliburn Micro works, please visit the project's website at `http://www.caliburnproject.org/`.

If you run the solution, the projects will be built and the following output will be shown:

Summary

In this chapter, we took a look at all the project templates available for Windows Phone Application development with F#. All three project templates had almost the same composition. Each template had the AppHost and App projects. AppHost is the Silverlight Application needed for the startup, and all the application logic and UI resides in App. We looked at three main examples for Windows Phone App, namely, Basic application, List application, and Panorama application.

In the next chapter, we will learn how to work with the Windows Phone controls. We will also learn to interact with most of the controls, that is, on click or selection changed, and so on.

3
Working with Windows Phone Controls

In the previous chapter, we familiarized ourselves with the different project templates available for F# and Windows Phone application development. We looked at three project templates that are available from the F# community and that help us in getting started with the development.

Silverlight runtime for Windows Phone provides a variety of controls as part of the framework itself. This chapter is all about getting to know the controls supported by the framework and understanding how to work with those controls. We will go through each control, understand its usage, and finally learn how to write code to work with those controls.

Supported controls in Windows Phone

The following list will illustrate the different controls supported in Windows Phone. These controls are included in the `System.Windows.Controls` namespace in the .NET Framework class library for Silverlight:

- `Button`: As the name goes, this is a button wherein a user interacts by clicking on it. On clicking, it raises an event.

- `HyperlinkButton`: This is a button control that displays a hyperlink. When clicked, it allows users to navigate to an external web page or content within the same application.

- `ProgressBar`: This is a control that indicates the progress of an operation.

- `MessageBox`: This is a control that is used to display a message to the user and optionally prompts for a response.

- `TextBox`: This is a control that allows users to enter single or multiple lines of text.

- `Checkbox`: This is a control that a user can select or clear, that is, the control can be checked or unchecked.

- `ListBox`: This is a control that contains a list of selectable items.

- `PasswordBox`: This is a control used for entering passwords.

- `RadioButton`: This is a control that allows users to select one option from a group of options.

- `Slider`: This is a control that allows users to select from a range of values by moving a thumb control along a track.

Hello world in F#

The previous section gave us an insight into different controls available for Windows Phone applications. Before understanding how to work with them, let's create a Windows Phone "Hello World" application using F#. The following steps will help us create the application:

1. Create a new project of type **F# and C# Windows Phone Application (Silverlight)** (refer to *Chapter 2, F# Windows Phone Project Overview* to know more about the F# Windows Phone project template types). A solution with **App** and **AppHost** projects will be created:

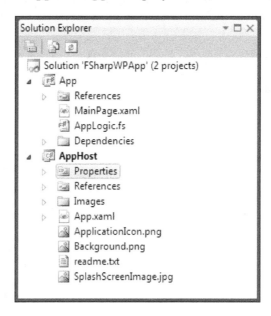

2. In the **App** project, we will have the main visual for the application called
   `MainPage.xaml`. If you open `MainPage.xaml`, you will notice that `MainPage`
   is actually a `PhoneApplicationPage` type. This is evident from the following
   XAML declaration:

```
<phone:PhoneApplicationPage
x:Class="WindowsPhoneApp.MainPage"
xmlns="http://schemas.microsoft.com/winfx/2006/xaml/presentation"
xmlns:x="http://schemas.microsoft.com/winfx/2006/xaml"
xmlns:phone="clr-namespace:Microsoft.Phone.
Controls;assembly=Microsoft.Phone"
xmlns:shell="clr-namespace:Microsoft.Phone.
Shell;assembly=Microsoft.Phone"
xmlns:system="clr-namespace:System;assembly=mscorlib"
xmlns:swc="clr-namespace:System.Windows.Controls;assembly=System.
Windows"
xmlns:d="http://schemas.microsoft.com/expression/blend/2008"
xmlns:mc="http://schemas.openxmlformats.org/markup-
compatibility/2006"
FontFamily="{StaticResource PhoneFontFamilyNormal}"
FontSize="{StaticResource PhoneFontSizeNormal}"
Foreground="{StaticResource PhoneForegroundBrush}"
SupportedOrientations="Portrait" Orientation="Portrait"
shell:SystemTray.IsVisible="True" mc:Ignorable="d"
d:DesignHeight="696" d:DesignWidth="480">
```

Note the `x:Class` attribute; this denotes that the XAML contains a
counterpart class called `MainPage` available in the `WindowsPhoneApp`
namespace. The `MainPage` class can be found in the `AppLogic.fs` file
in the **App** project.

3. Let us take a closer look at the UI itself. The main contents of the application
   is contained in a grid. A grid is a layout control that is used to define a flexible
   area that consists of rows and columns. The body contains three `TextBlock`
   controls. A `TextBlock` control, as the name suggests, is used to display a
   small block of text. We have three `TextBlock` controls on the page, one for
   `ApplicationTitle`, another for `PageTitle`, and the last one for `Results`.
   There is also an empty grid named `ContentGrid`. So this is where we will
   be creating our "Hello World" experiment. The XAML for the content is
   shown as follows:

```
<Grid x:Name="LayoutRoot" Background="Transparent">
<Grid.RowDefinitions>
    <RowDefinition Height="Auto"/>
```

```
        <RowDefinition Height="*"/>
    </Grid.RowDefinitions>
    <StackPanel x:Name="TitlePanel" Grid.Row="0" Margin="24,24,0,12">
    <TextBlock x:Name="ApplicationTitle" Text="AN F# APPLICATION"
    Style="{StaticResource PhoneTextNormalStyle}"/>
    <TextBlock x:Name="PageTitle" Text="main page" Margin="-3,-8,0,0"
    Style="{StaticResource PhoneTextTitle1Style}"/>
    <TextBlock x:Name="Results" Text="" Margin="-3,-8,0,0"
     Style="{StaticResource PhoneTextTitle1Style}"/>
    </StackPanel>
    <!--ContentPanel - place additional content here-->
        <Grid x:Name="ContentGrid" Grid.Row="1">
        </Grid>
    </Grid>
```

As you can see from the code, ContentGrid is empty. So let's place a TextBlock control and a Button element inside ContentGrid. The idea is to generate the text "Hello World" when we click on the button.

4. First, let's take a look at the XAML portion of the "Hello World" experiment in MainPage.xaml:

```
<Grid x:Name="ContentGrid" Grid.Row="1">
  <Grid.RowDefinitions>
    <RowDefinition Height="*"/>
      <RowDefinition Height="*"/>
  </Grid.RowDefinitions>
   <Grid.ColumnDefinitions>
    <ColumnDefinition Width="*" />
   </Grid.ColumnDefinitions>
   <TextBlock  Grid.Row="0" Grid.Column="0"
              Name="txtMessage"
              Text="Click the button"
              HorizontalAlignment="Center"
              VerticalAlignment="Center"
              Style="{StaticResource PhoneTextTitle2Style}"/>
  <Button  Grid.Row="1" Grid.Column="0"
      Name="btnSayHelloButton"
          Height="100" Content="Click to say Hello !" />
</Grid>
```

Pay attention to the TextBlock element and the Button names. We have the TextBlock control named as txtMessage and Button named as btnSayHelloButton.

5.   Now the second part of this experiment is to wire up the button's `Click` event with an event handler in the `MainPage` class. In the `AppLogic.fs` file, find the `MainPage` type and add the following code:

```
// Bind named Xaml components relevant to this page.
let txtMessage : TextBlock
   = this?txtMessage
let btnSayHelloButton : Button
   = this?btnSayHelloButton
//Wire Click Event
do btnSayHelloButton.Click.Add(fun _ ->
   txtMessage.Text <- "Hello World !")
```

6.   First we create a reference to the text block and the button. Then we add an event handler to the button's `Click` event. In F#, the way we add event handlers is by writing a function using the `fun` keyword. The _ (underscore) tells the F# compiler to ignore the parameters of the function and then we define the statement of a function. On button click, we just change the text of the text block to say "Hello World !". Well, that's all there is to this "Hello World" experiment.

 Notice the use of the ? operator. This is not F#-specific code. Rather, the project template creates a module in the `AppLogic.fs` file called `Utilities`. There, ? is defined as an operator that can be used for dynamic lookup of XAML object names for binding purposes.

The code snippet of the operator is shown as follows:

```
/// This is an implementation of the dynamic lookup operator
f//or binding Xaml objects by name.
let (?) (source:obj) (s:string) =
      match source with
   | :? ResourceDictionary as r ->   r.[s] :?> 'T
   | :? Control as source ->
      match source.FindName(s) with
      | null -> invalidOp (sprintf "dynamic lookup of Xaml
component %s failed" s)
      | :? 'T as x -> x
      | _ -> invalidOp (sprintf "dynamic lookup of Xaml component
%s failed because the component found was of type %A instead of
type %A"  s (s.GetType()) typeof<'T>)
```

```
    |  _ -> invalidOp (sprintf "dynamic lookup of Xaml component
%s failed because the source object was of type %A. It must be a
control or a resource dictionary" s (source.GetType()))
```

7. Now let's build and run the project. Windows Phone Emulator will be invoked by Visual Studio to deploy the app we just built. You will see a text block with the text **Click the button** and a button with the text **Click to Say Hello !**. When the button is clicked, the text block will show the text **Hello World !**. The screenshots of the final output are shown as follows:

Working with the Button control

A button, as the name goes, is a rectangular control that allows a user to click on it, and when clicked, raises a Click event. We can write listeners for this Click event and add an event handler for the Click event. When the Click event occurs, the event handler will be notified and we can run our business logic against the button click—whatever logical thing we need. Let's see how to work with the button control.

Create a project and add three buttons in the XAML code. For the first button, we will set its properties from the XAML itself. For the second button, we will set the properties from the code. For the third button we will set its properties in its Click event. The XAML code snippet is shown as follows:

```
<Button  Grid.Row="0"
         Name="btnFirstButton"
```

```
        Content="First Button"
        Width="300"
        VerticalAlignment="Center"
        HorizontalAlignment="Center"
        Background="Orange"
        Foreground="Black"/>
<Button  Grid.Row="1"
        Name="btnSecondButton"
        Content="Second Button"
        />
<Button  Grid.Row="2"
        Name="btnThirdButton"
        Content="Third Button"
        />
```

For the second and third button, except for its Content attribute, nothing is set in XAML. The properties for the second button is set on the page load event in the MainPage class. The properties for the third button is set on the click of the third button in an event handler. Now let us see the F# code snippet for this in the MainPage class:

```
//Handle Page Loaded event to set properties for Second Button
do this.Loaded.Add(fun _ ->
  btnSecondButton.Width <- 300.0
    btnSecondButton.Height <- btnFirstButton.ActualHeight
    btnSecondButton.Content <- "Runtime Text"
    btnSecondButton.Background <- greenColorBrush
    btnSecondButton.Foreground <- yellowColorBrush
)
//Handle Click event on Third button to set properties
do btnThirdButton.Click.Add(fun _ ->
    btnThirdButton.Width <- 300.0
    btnThirdButton.Height <- btnFirstButton.ActualHeight
    btnThirdButton.Content <- "Button Clicked"
    btnThirdButton.Background <- redColorBrush
    btnThirdButton.Foreground <- whiteColorBrush
  btnThirdButton.VerticalAlignment <-        VerticalAlignment.Center
    btnThirdButton.HorizontalAlignment <-      HorizontalAlignment.
Center
  )
```

One thing to learn from here is—whatever properties can be set from XAML, the same can also be set from the code. The preceding demo shows how at page load and with event handlers, a control's properties can be changed at runtime. The screenshot of the final output is shown as follows:

Working with the Checkbox control

As mentioned earlier, Checkbox is a control that allows a user to select or clear an option. We can use a Checkbox control to provide a list of options that a user can select, for example a list of settings to apply in an application. The Checkbox control can have three states namely Checked, Unchecked, and Indeterminate.

To demonstrate this control usage, let's build a demo that contains two checkboxes. The first checkbox demonstrates the Checked and Unchecked states. The second checkbox demonstrates the Checked, Unchecked, and Indeterminate states. We will handle the Checked event when checkboxes are checked, and the Unchecked event when checkboxes are unchecked. The XAML code snippet for this demo is shown as follows:

```
<Grid x:Name="ContentGrid" Grid.Row="1">
  <StackPanel>
    <!--Two State Checkbox-->
    <CheckBox x:Name="chkBox1" Content="Two State" />
```

```
        <!--Three State Checkbox-->
        <CheckBox x:Name="chkBox2"
                  IsThreeState="True"
                  Content="Three State" />
        <!-- Message text block -->
        <TextBlock x:Name="txtMessage" />
    </StackPanel>
</Grid>
```

As you can see, we have two checkboxes stacked vertically one below the other. StackPanel is a layout control, which, as its name goes, just stacks its children content either vertically or horizontally. The second checkbox has a Boolean property named IsThreeState set to true. That means this checkbox will have three states – Checked, Unchecked, and Indeterminate. Checkboxes expose Checked, Unchecked, and Indeterminate events. We will wire up event handlers for these events and write out a message to the txtMessage text block as seen in the code snippet. The following is the code snippet where we handle the events:

```
let checkBoxOne : CheckBox = this?chkBox1
let checkBoxTwo : CheckBox = this?chkBox2
let txtMessage : TextBlock = this?txtMessage

let UpdateMessage message =
  txtMessage.Text <- message

do checkBoxOne.Checked.Add(fun _ ->
  do UpdateMessage "Two State Checkbox Checked"
)
do checkBoxOne.Unchecked.Add(fun _ ->
  do UpdateMessage "Two State Checkbox UnChecked"
)
do checkBoxTwo.Checked.Add(fun _ ->
  do UpdateMessage "Three State Checkbox Checked"
)
do checkBoxTwo.Unchecked.Add(fun _ ->
  do UpdateMessage "Three State Checkbox UnChecked"
)
do checkBoxTwo.Indeterminate.Add(fun _ ->
  do UpdateMessage "Three State Checkbox Indeterminate"
)
```

We first get a reference to the checkbox controls. Then we wire up the `Checked` and `Unchecked` events. For the second checkbox, since it supports the `Indeterminate` state, we wire up the `Indeterminate` event too. When you run the app and select or clear any checkbox, a message will be shown in the text block. The screenshot of the output is shown as follows:

Working with the Hyperlink control

`Hyperlink` is a control that presents a button control with a hyperlink. When the hyperlink is clicked, it will navigate to the URI specified, which can be an external web page or content within the app. We specify the URI to navigate through the `NavigateUri` property. The XAML code snippet for this control is shown as follows:

```
<HyperlinkButton
            Content="Click here to learn about Silverlight"
            NavigateUri=http://www.silverlight.net
            TargetName="_blank" />
```

The same effect can be obtained using code. On page load, we would have to just set the `NavigateUri` property, and when the user clicks on the hyperlink button, he will be navigated to the set URI.

Working with the ListBox control

A `ListBox` control represents a list of selectable items. It basically displays a collection of items. More than one item in a `ListBox` control is visible at a time.

As part of the demo app, we will create a listbox and fill it with available color names. When an item is selected in the listbox, we will set the background of the listbox to the selected item. The XAML code snippet is shown as follows:

```
<Grid x:Name="ContentGrid" Grid.Row="1">
  <ListBox x:Name="lstColors" />
</Grid>
```

The code to fill up the listbox with the names of the colors along with the event handler to handle the listbox's `SelectionChanged` event is shown as follows:

```
let colorsListBox:ListBox = this?lstColors
// Fill ListBox with Color names
let props = (typeof<Colors>).GetProperties()
do props |> Seq.iter (fun p -  > colorsListBox.Items.Add(p.
Name)|>ignore)
// ListBoxOnSelectionChanged
do colorsListBox.SelectionChanged.Add( fun _ ->
  let str = colorsListBox.SelectedItem :?> string
   if str <> null then
  let clr = (typeof<Colors>).GetProperty(str).
GetValue(null,null) :?> Color
    colorsListBox.Background <- new SolidColorBrush(clr)
)
```

For filling up the listbox with color names, we iterate through the public properties of the `System.Windows.Media.Colors` class. The `Colors` class implements a different set of predefined colors. We fill the listbox with the names of the predefined colors by adding them to the `Items` collection of the listbox.

To handle item selection change, we handle the `SelectionChanged` event. First, we get the `SelectedItem` property, and since we know it's a string in our case, we convert it into a string. Then we get the `Color` property by making use of the string that we converted from `SelectedItem`. Once we get the color, we set the background of the listbox to the color selected.

The final output of this demo is shown as follows:

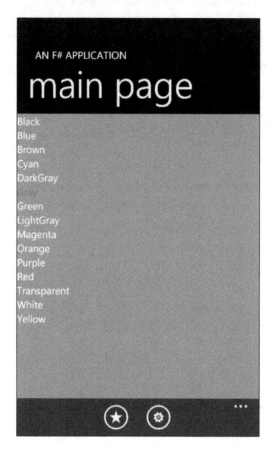

Working with the MessageBox control

In this section we will take a look at the MessageBox control. This control displays a message to the user and optionally prompts for a response. The MessageBox class provides a static Show method, which can be used to display the message in a simple dialog box. The dialog box is modal and provides an OK button.

A code to work with the MessageBox control is shown next. Note that this can be worked with only from the code and not from the XAML. First, we show a message box with the **ok** and **cancel** button. When a user clicks on the **ok** button, we show a simple message box with just the **ok** button.

```
let mutable messageBoxResult:MessageBoxResult
  = MessageBoxResult.None
let messageBody = "Would you like to see the simple version?"
let messageCaption = "MessageBox Example"

do messageBoxResult <- MessageBox.Show(
  messageBody,
  messageCaption,
  MessageBoxButton.OKCancel)

if(messageBoxResult = MessageBoxResult.OK) then
  MessageBox.Show("No caption, one button.") |> ignore
```

The final output of this demo is shown as follows:

Working with the PasswordBox control

PasswordBox, as the name suggests, is used to enter a password in applications. The user cannot view the entered text; only password characters that represent the text are displayed. The password character to be displayed can be specified by using the property PassowrdChar.

Add `PasswordBox`, `Button`, and `TextBlock` in the XAML code. The idea is to enter some text in the PasswordBox control, and on clicking the button show the password text in the text block. The XAML for this demo is shown as follows:

```
<Grid x:Name="ContentGrid" Grid.Row="1">
  <StackPanel>
    <PasswordBox x:Name="pwdBox" MaxLength="8"
    <Button x:Name="btnShowPassword"
Content="Click to show Password" />
    <TextBlock x:Name="txtPassword"/>
  </StackPanel>
</Grid>
```

The code to handle the button click and display the password entered in the text block is shown as follows:

```
let passwordBox : PasswordBox = this?pwdBox
 let passwordShowButton : Button = this?btnShowPassword
let passwordTextBlock : TextBlock = this?txtPassword

    do passwordShowButton.Click.Add(fun _ ->
        passwordTextBlock.Text <- passwordBox.Password
    )
```

The password box contains a property called `Password`, which can be used to read the entered password. The final output of the demo is shown as follows:

Working with the ProgressBar control

The `ProgressBar` control is used to display the progress of an operation. This is often used in UI layouts to indicate a long running operation. One of the requirements of the Windows Phone app is to include a progress bar and show a progress animation whenever a task is a long-running task in any application. The progress bar can have a range between `Minimum` and `Maximum` values. It also has an `IsIndeterminate` property, which means no `Minimum` and `Maximum` value is set and the progress bar displays a repeating pattern. This is predominantly used in XAML and its visibility is controlled by the code. The XAML code snippet is shown as follows:

```
<ProgressBar  x:Name="pg1" Value="100"  Margin="10"
              Maximum="200" Height="15"
              IsIndeterminate="False" />

<ProgressBar  x:Name="pg2" Margin="10" Height="15"
              IsIndeterminate="True"    />
```

Working with the RadioButton control

`RadioButton` is a control, which represents a button that allows a user to select a single option from a group of options. A `RadioButton` control is usually used as one item in a group of `RadioButton` controls. `RadioButtons` can be grouped by setting their `GroupName` property. To group radio buttons, the `GroupName` property on each of the radio button should have the same value. `RadioButtons` contain two states, namely selected or cleared. `RadioButtons` have the `IsSelected` property, which will let us know if a radio button is selected or not.

Create three radio buttons in XAML. Two of them will be grouped and one will be ungrouped. We will listen for the `Checked` event on the radio buttons and update a text block with the appropriate message. The XAML code snippet is shown as follows:

```
<Grid x:Name="ContentGrid" Grid.Row="1" Margin="12">
  <StackPanel>
    <TextBlock Text="First Group:"  Margin="5" />
    <RadioButton   x:Name="TopButton"
                   Margin="5"
                   GroupName="First Group"
                   Content="First Choice" />
    <RadioButton   x:Name="MiddleButton"
                   Margin="5"
```

```
                    GroupName="First Group"
                    Content="Second Choice" />
    <TextBlock Text="Ungrouped:" Margin="5" />
      <RadioButton    x:Name="LowerButton"
                    Margin="5"
                    Content="Third Choice" />
    <TextBlock x:Name="choiceTextBlock" Margin="5" />
  </StackPanel>
</Grid>
```

As you can see, the first two radio buttons have their GroupName property set whereas the last radio button does not have any GroupName set. We will wire up the Checked event on all three radio buttons and update the text block with information such as which radio button was clicked. The code snippet is shown as follows:

```
let topButton : RadioButton = this?TopButton
let middleButton : RadioButton = this?MiddleButton
let lowerButton : RadioButton = this?LowerButton
let choiceTextBlock : TextBlock = this?choiceTextBlock

let UpdateMessage (radioButton : RadioButton) =
    choiceTextBlock.Text <- "You Chose: " +
    radioButton.GroupName + " " +
      radioButton.Name
do topButton.Checked.Add(fun _ ->
    do UpdateMessage topButton
)
do middleButton.Checked.Add(fun _ ->
    do UpdateMessage middleButton
)
do lowerButton.Checked.Add(fun _ ->
    do UpdateMessage lowerButton
)
```

The output from this demo is shown as follows:

Working with the Slider control

The Slider control represents a control that lets users select from a range of values by moving a thumb control along a track.

The Slider control exposes certain properties that can be set to customize the functioning of the slider. We can set the Orientation property to orient the slider either horizontally or vertically. We can change the direction of the increasing value with IsDirectionReversed. The range of values can be set using the Minimum and Maximum properties. The value property can be used to set the current position of the slider.

Add a Slider control to the XAML. Set its Minimum to 0 and Maximum to 10. When the user changes the position of the thumb on the slider, we will listen to the ValueChanged event on the slider and show the current value in a text block. The XAML snippet for the slider is shown as follows:

```
<TextBlock Text="Slider with ValueChanged event handler:" />
<Slider    Margin="0,5,0,0"
           x:Name="slider"
```

```
            Minimum="0"
            Maximum="10"
        />
<TextBlock  Margin="0,5,0,20"
            x:Name="txtMessage"
            Text="Current value: 0" />
```

The code snippet is shown as follows:

```
let slider : Slider = this?slider
let txtMessage : TextBlock = this?txtMessage

do slider.ValueChanged.Add(fun _ ->
  txtMessage.Text <-
  "Current value: " + slider.Value.ToString()
)
```

As you can see, we set the `Minimum` and `Maximum` range in the XAML. From the code, we wire up the `ValueChanged` event. Whenever a user changes the value using the thumb on the slider, the `ValueChanged` event will be fired and we just read the current value of the slider and update a text block. The final output of this demo is shown as follows:

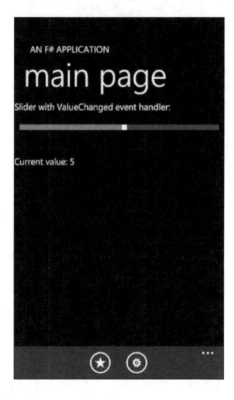

Working with the TextBox control

The `TextBox` control can be used to display single or multiline text. It is often used to accept user input in applications. This control is one of the most widely used controls for data input.

On a Windows Phone, whenever a textbox gets focus, an on-screen keyboard known as **Software Input Panel** (**SIP**) will be shown automatically by the Windows Phone OS. If we do not want the user to edit the text, we can set the `IsReadOnly` property on the textbox to `true`. This will prevent the user from typing anything in the textbox. We can read the value entered in a textbox using the `Text` property. The XAML snippet for a simple textbox is shown as follows:

```
<TextBox x:Name="ReadOnlyTB"
    IsReadOnly="True"
    HorizontalAlignment="Left"
    Height="35" Width="200" />
```

A screenshot of a simple textbox with SIP displayed when the textbox gets focus is shown as follows:

Summary

In this chapter, we took a lap around the supported controls for the Silverlight runtime on the Windows Phone platform. We looked at the XAML way of defining the controls and also how to programmatically work with these controls in the code. We learnt what properties each control exposes and how to wire up events supported by each control. This gives us a head start for the coming chapters where we will start looking at other aspects related to Windows Phone.

4
Windows Phone Screen Orientations

In the previous chapter, we looked into all the supported controls of the Windows Phone Silverlight runtime. We learned about the different types of controls and how to work with them in F#. In this chapter, we will look into one of the interesting concepts of Windows Phone – orientation. Since we are dealing with a handheld device, the device itself can be held either vertically or horizontally. In technical terms, this tilting of the device is termed as orientation. This chapter will explain the orientation behavior of the device and how, as a developer, you should handle the orientation changes in your application code.

Orientation

With Windows Phone being a handheld device, the user has all the freedom to turn the device upside down or rotate the device to the left or right. The behavior of turning the phone upside down or rotating it to the left or right is technically termed as orientation in Windows Phone development. As a Windows Phone application developer, you are supposed to take care of the orientation effects in the app you develop. You need to decide whether you would like your app to react to the orientation changes or not. To provide a better user experience, it is a best practice to include the orientation aspect while designing the app.

Types of orientations

Windows Phone supports three possible orientations on the device. They are:

- Portrait
- Landscape (left)
- Landscape (right)

Portrait orientation is nothing but the vertical positioning of the content, that is, when the phone is held straight up. In the portrait orientation, the screen resolution is 480 x 800 px. 480 x 800 px is a standard screen resolution on Windows Phone devices. That means we get to work with a standardized resolution set when designing for Windows Phone.

A screenshot of how the portrait orientation looks with respect to a physical device is shown as follows:

Portrait

Landscape orientation is the horizontal positioning of the content, that is, when the phone is rotated to the left or right. A user can cause the change in orientation by just rotating the phone. In this orientation mode, the width (800 px) will be greater than height (480 px). As mentioned earlier, we can have landscape left and landscape right. The difference between these two modes is on which side the status bar will be after rotating the device horizontally. Landscape left means that the status bar is found to the user's left and landscape right means that the status bar is found to the user's right.

Landscape right and landscape left

Setting orientation

The orientation can be specified at a single-page level. One advantage of specifying the orientation at single-page level is, we can design different pages supporting different orientations.

To specify the supported orientation by a page, we set the `SupportedOrientation` property on the XAML. The code for setting `SupportedOrientation` is shown as follows:

```
SupportedOrientations="PortraitOrLandscape"
```

`SupportedOrientation` is a property on the `PhoneApplicationPage` type – which is usually the page we are designing. `SupportedOrientations` is an enum of type `SupportedPageOrientation`. `SupportedPageOrientation` has the following items as part of the enumeration:

- `Portrait` – portrait orientation
- `Landscape` – landscape orientation; landscape supports both left and right views, but there is no way, programmatically, to specify one or the other
- `PortraitOrLandscape` – landscape or portrait orientation

In order for a page to change its orientation, the SupportedOrientations property should be set to one of the three available values. If the property is set to Portrait, when the user rotates the phone to the landscape mode, the page will not be rotated to be in the landscape mode. So this property controls what orientation a page supports.

Detecting orientation

Since the orientation is not controlled by an application but by the user, the Windows Phone application pages have a property that will let us know what orientation the phone has at any moment. The phone application page also allows us to listen to the OrientationChanged event and react accordingly. In order to detect an orientation change, we can wire up the OrientationChanged event and provide logic to handle the orientation change. The code snippet is shown as follows:

```
do this.OrientationChanged.Add(
    fun (args : OrientationChangedEventArgs)  ->
    if(args.Orientation = PageOrientation.Portrait)
    then
            (
            )
        else
            (
            )
)
```

Changing orientation in an emulator

When you are developing an application, you may wonder how to test the orientation effects on your application. Well, the answer to this is the Windows Phone emulator. The emulator provides options to rotate the emulator either as a portrait or as a landscape. So using this feature you can test whether your app is reacting correctly to the orientation changes.

The following screenshot shows the emulator buttons, which are also known as the orientation buttons. The orientation buttons have rectangles with arrows that indicate the orientation change:

Orientation handling techniques

In the previous sections we learned how to support and detect orientation. The next question would be, what techniques are available to handle orientation in our application? Well, the answer to that is, there are two mechanisms for displaying content in the portrait or landscape mode. They are:

- **Auto sizing and scrolling** – this mechanism makes use of StackPanel and ScrollViewer to support both orientations. It is a simple and good mechanism to use when the content appears in a list or if different controls appear one after another on the page.

- **Grid layout** – this mechanism makes use of a grid to place the controls and the position of the elements is moved within the grid cells, based on the orientation change.

Auto sizing and scrolling

This mechanism uses a StackPanel control placed within a ScrollViewer control. StackPanel, as the name goes, will enable us to order the child controls one after the other. The ScrollViewer control enables us to scroll through the StackPanel control if the UI elements don't fit on screen when we switch the orientation from portrait to landscape. The following steps help in implementing this mechanism:

1. The SupportedOrientations property on the page needs to be set to PortraitOrLandscape.

2. Grid, which is used as the content layout on the page, needs to be replaced with ScrollViewer and StackPanel.

The XAML snippet is shown as follows:

```
<Grid x:Name="LayoutRoot" Background="Transparent">
    <Grid.RowDefinitions>
        <RowDefinition Height="Auto"/>
        <RowDefinition Height="*"/>
    </Grid.RowDefinitions>
    <!--TitlePanel contains the name of the application and page
title-->
    <StackPanel x:Name="TitlePanel" Grid.Row="0"
Margin="24,24,0,12">
        <TextBlock x:Name="ApplicationTitle" Text="AN F#
APPLICATION"
                Style="{StaticResource PhoneTextNormalStyle}"/>
        <TextBlock x:Name="PageTitle" Text="main page"
Margin="-3,-8,0,0"
                Style="{StaticResource PhoneTextTitle1Style}"/>
        <TextBlock x:Name="Results" Text="" Margin="-3,-8,0,0"
                Style="{StaticResource PhoneTextTitle1Style}"/>
    </StackPanel>
    <!--ContentPanel - place additional content here-->
    <ScrollViewer x:Name="ContentGrid" Grid.Row="1" VerticalScrollB
arVisibility="Auto">
        <StackPanel>
            <Border Height="80" Background="Gray" Margin="12">
                <TextBlock VerticalAlignment="Center">Line 1</
TextBlock>
            </Border>
            <Rectangle Width="100" Height="100" Margin="12,0"
                HorizontalAlignment="Left" Fill="{StaticResource
PhoneAccentBrush}"/>
            <Rectangle Width="100" Height="100"
HorizontalAlignment="Center"
                Fill="{StaticResource PhoneAccentBrush}"/>
            <Rectangle Width="100" Height="100" Margin="12,0"
                HorizontalAlignment="Right" Fill="{StaticResource
PhoneAccentBrush}"/>
            <Border Height="80" Background="Gray" Margin="12">
                <TextBlock VerticalAlignment="Center">Line 5</
TextBlock>
```

```
        </Border>
        <Border Height="80" Background="Gray" Margin="12">
            <TextBlock VerticalAlignment="Center">Line 6</
TextBlock>
        </Border>
      </StackPanel>
    </ScrollViewer>
  </Grid>
```

As you can see, we have used the `ScrollViewer` control instead of a `Grid` control as the content holder. Then we have a `StackPanel` control as the child of the `ScrollViewer` control. The `StackPanel` control contains three `TextBlock` classes and three `Rectangle` classes stacked one below the other. We have set the `SupportedOrientations` property on the page to `PortraitOrLandscape`. The screenshot of the output in the portrait mode is shown as follows:

The output in the landscape mode is shown as follows:

As you can see, this technique relies on XAML alone and no code is required to react to the orientation change. In the next section, we will learn about the grid layout mechanism for handling the orientation.

Grid layout

In this mechanism, we position the UI elements in a grid. We handle the orientation change through code by programmatically repositioning the UI elements in the grid into different cells. In this mechanism too, we set the SupportedOrientations property of the page to PortraitOrLandscape. We let Grid be the layout for the content of the page. We hook into the OrientationChanged event, create an event handler, and add code to reposition the UI elements within the grid.

As a demo to illustrate this mechanism, we will create a page with a grid that has two rows and two columns. We will position a rectangle in the first row and first column and a set of buttons in the second row and first column. We will then add an event handler for OrientationChanged and reposition the elements.

The XAML snippet is shown as follows:

```
<Grid x:Name="ContentGrid" Grid.Row="1">
        <Grid.RowDefinitions>
            <RowDefinition Height="Auto"/>
            <RowDefinition Height="*"/>
        </Grid.RowDefinitions>
        <Grid.ColumnDefinitions>
            <ColumnDefinition Width="Auto"/>
```

```
            <ColumnDefinition Width="*"/>
        </Grid.ColumnDefinitions>

        <Border x:Name="myRectangle"
                Background="Gray"
                Grid.Row="0" Grid.Column="0"
                HorizontalAlignment="Center"
                Height="300" Width="500">
            <TextBlock HorizontalAlignment="Center"
                    VerticalAlignment="Center"
                    Text="Rectangle" />
        </Border>
        <StackPanel x:Name="buttonList"
                Grid.Row="1" Grid.Column="0"
                HorizontalAlignment="Center" >
            <Button Content="Button 1" />
            <Button Content="Button 2" />
            <Button Content="Button 3" />
            <Button Content="Button 4" />
        </StackPanel>
    </Grid>
```

The code which handles the OrientationChanged event is shown as follows:

```
    let buttonList : StackPanel = this?buttonList
    do this.OrientationChanged.Add(
        fun (args : OrientationChangedEventArgs)  ->
            if(args.Orientation = PageOrientation.Portrait) then
                (
                    Grid.SetRow(buttonList, 1)
                    Grid.SetColumn(buttonList, 0)
                )
            else
                (
                    Grid.SetRow(buttonList, 0)
                    Grid.SetColumn(buttonList, 1)
                )
    )
```

As you can see in the code, we check for the orientation. If it is `Portrait`, we position the `buttonList` to be in the second row and first column. If the orientation is `Landscape`, we position the `buttonList` to the first row, second column. The screenshot of the output when in the portrait mode is shown as follows:

The screenshot of the output when in the landscape mode is shown as follows:

Summary

In this chapter we looked into one of the main aspects of the Windows Phone device, namely, orientation. We understood what orientation is, what are the types of orientations supported by the device, how to detect an orientation, and the techniques needed to deal with the orientation changes.

5
Windows Phone Gesture Events

In this chapter, we will look at one of the unique aspects of the Windows Phone platform, namely, the gesture awareness and gesture support it provides for application and game development. We will look at the following topics in this chapter:

- Gestures
- Gesture events
- Handling gestures

Gestures

Most of the smartphones in the present era provide the ability to touch the screen when making a choice or selecting an option. Gone are the days when phones had only hardware keyboards and we had to use keys to navigate around the choices and options in order to pick one. Present day smartphones have the ability to detect single touch, that is, when you use one finger to touch the screen and select an option or multitouch, that is, using more than one finger to touch the screen.

Windows Phone devices are capable of detecting multitouch as they all have the requirement to support multitouch screens. That means you as a user can use multiple fingers to produce different inputs, particularly input gestures like tapping, flicking, or pinching. For a user, touch gestures are the primary way to interact with a Windows Phone. Tapping on a button or listbox item is one such example of an input gesture on the phone.

Gesture support in Silverlight for Windows Phone

The controls supported by Silverlight for Windows Phone are gesture aware, that is, they understand gesture inputs when used in an application. Gesture inputs include single finger touch, multi-finger touch, tapping, holding, and swiping on the screen. We can process the touch input by making use of events supported by the platform. Also, almost all the visual elements allow us to handle simple gesture events such as tap, double tap, tap, and hold by making use of the events available on the UI element.

Manipulation events

Manipulation events on visual elements allow us to move and scale objects for touch or multitouch input. Here is the description of the manipulation events available in Silverlight for Windows Phone:

Event	Description
ManipulationStarted	Occurs when the user starts a direct manipulation by placing his/her finger or fingers on the screen
ManipulationDelta	Occurs repeatedly while the user is moving his/her finger or fingers on the screen
ManipulationCompleted	Occurs when the user removes his/her finger or fingers from the screen

Gesture events

As stated earlier all visual elements of Silverlight for Windows Phone are gesture-aware and support simple gesture events such as tap, double tap, and tap and hold. These simple gesture events are supported as events on UI elements. Here is the description of the gesture events in Silverlight for Windows Phone:

Event	Description
Tap	A finger touches the screen and releases
DoubleTap	Two taps in succession represents a double tap
Hold	A finger touches the screen and holds it in place for a brief period of time

Understanding manipulation events

As seen earlier, the manipulation event in Silverlight for Windows Phone allows us to handle certain gesture events to move and scale a UI element. The order in which these manipulation events are fired is – ManipulationStarted, ManipulationDelta, and ManipulationCompleted. Let us understand these events one by one:

- ManipulationStarted – this event is fired when a user touches the screen.
- ManipulationDelta – after the ManipulationStarted event has been fired, the ManipulationDelta event is fired one or more times. Consider a situation where you touch the screen and then drag your finger across the screen. The ManipulationDelta event is fired when you drag.
- ManipulationCompleted – this event is fired when the gesture is completed. For example, when you release your finger from the screen, it is assumed that the gesture is completed.

Let's consider a simple demo to understand the concepts of manipulation events. What we will do is, we will place a rectangle on the screen and users can use their fingers to hold the rectangle and move it across the screen. We will set up the ManipulationDelta event to handle the drag gesture on the rectangle.

First, let's create a rectangle with the Width and Height attribute set to 150. We will fill the rectangle with a Gray color:

```
<Rectangle Name="myRectangle"
           Width="150"
           Height="150"
           Fill="Gray" />
```

Next we make use of the TranslateTransform object by creating a new instance of it, to translate the rectangle. TranslateTransform is a class which moves an object in two dimensional X-Y coordinate systems. We then set up the ManipulationDelta event handler on the rectangle. We provide the TranslateTransform object we created to the rectangle's RenderTransform. In the ManipulationDelta event handler, we just update the position of the rectangle by using the TranslateTransform object.

Here is the code snippet:

```
let dragTransTransform : TranslateTransform =
      new TranslateTransform()
  let myRectangle:Rectangle =
      this?myRectangle
```

```
do myRectangle.RenderTransform <-
    dragTransTransform
do myRectangle.ManipulationDelta.Add(
    fun (args : ManipulationDeltaEventArgs) ->
        do dragTransTransform.X <-
            dragTransTransform.X + args.DeltaManipulation.
Translation.X
        do dragTransTransform.Y <-
            dragTransTransform.Y + args.DeltaManipulation.
Translation.Y
    )
```

As you can see, we first create an object of `TranslateTransform`. We then get a reference to our rectangle and assign the `TranslateTransform` object we created to the `RenderTransform` property of the rectangle. In the event handler for the `ManipulationDelta` event, we just get the new X-Y coordinate and add it to the previously held X-Y coordinates. This makes the rectangle move along the path in which the user moves his/her finger.

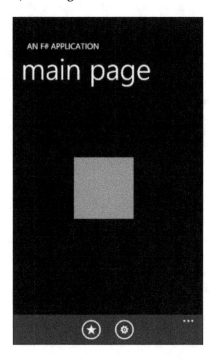

Next let us see the gesture events supported by Silverlight for Windows.

Understanding gesture events

Simple gestures like tap and double tap can be handled by using the mouse events. The mouse events are limited to simple gestures like tap and double tap only. A finger touch is converted to an equivalent mouse event. Some of the mouse events available in Silverlight for Windows Phone are as follows:

- `MouseLeftButtonDown` – event fired when a finger is placed on the screen
- `MouseLeftButtonUp` – event fired when a finger is lifted from the screen
- `MouseMove` – event fired when a finger is dragged across the screen
- `MouseLeave` and `MouseEnter` – as the name suggests, the event is fired when a finger enters the UI element and leaves the UI element

Let's create a simple demo to handle the `MouseLeftButtonUp`, `MouseLeftButtonDown`, and `MouseLeave` events, which showcases the tap gesture.

We will create a rectangle which is filled with `Red` color, with `Width` set to `200` and `Height` set to `100`. Here is the XAML snippet:

```
<Rectangle  x:Name="MyRectangle"
            Fill="Red"
            Height="100"  Width="200" />
```

Then we will handle the mouse events on the rectangle by setting up the event handlers. In the `MouseLeftButtonDown` event handler, we will increase the width and height of the rectangle. In `MouseLeftButtonUp` and `MouseLeave`, set the width and height of the rectangle back to the original values. Here is the code snippet:

```
let myRectangle:Rectangle = this?MyRectangle
    do myRectangle.MouseLeftButtonDown.Add(
        fun _ ->
            do myRectangle.Width <- 300.0
            do myRectangle.Height <- 200.0
    )
    do myRectangle.MouseLeftButtonUp.Add(
        fun _ ->
            do myRectangle.Width <- 200.0
            do myRectangle.Height <- 100.0
    )
    do myRectangle.MouseLeave.Add(
        fun _ ->
            do myRectangle.Width <- 200.0
            do myRectangle.Height <- 100.0
    )
```

Here is the screenshot of the output:

Summary

In this chapter, we took a lap around the different gestures supported by Silverlight for Windows Phone. We understood the different gesture events and how to work with them. We also saw how to use the manipulation and gesture events and work with them to either move or scale a UI element.

6
Windows Phone Navigation

In this chapter, we will look at one of the basic features for any application's development, namely navigation. We will look at the navigation support provided by Silverlight for Windows Phone. We will understand the different ways of performing page-to-page navigation and will also look at how to exchange data between pages.

Navigation in Windows Phone

Let us first try to understand the term navigation. Simply put, navigation allows a user to move through different pages of content within an application. Windows Phone applications are based on a page model, which is similar to that of Silverlight. Because of this, the navigation in Windows Phone applications is nothing but the presentation of different screens or different contents to the end user thereby allowing him to move back and forth between the content.

Since the Windows Phone hardware mandates a hardware back button, it can be used to go back one screen in any application. There is no need to provide an explicit navigation action to go back in Windows Phone applications.

This type of infrastructure allows the developers to achieve the following:

- Creation of different view-based or different screen-based content applications that fit naturally with that of the Windows Phone navigation model
- Easy to provide transition when navigating from one view to another that matches the default transition of the Windows Phone look and feel

Windows Phone navigation model

The navigation model in Windows Phone is based on one control called
`PhoneApplicationFrame`. `PhoneApplicationFrame` can contain one or more controls
named `PhoneApplicationPage`. The `PhoneApplicationPage` controls are the pages
or the screen through which a user can navigate. So `PhoneApplicationFrame` is
the main navigation control, which makes it possible to navigate to and from pages.
`PhoneApplicationPage`, as the name goes, represents a single page, which can
contain the content to be shown as part of that page. An application typically will
contain only one `PhoneApplicationFrame` and multiple `PhoneApplicationPage`.
A figure that depicts the relationship between `PhoneApplicationFrame` and
`PhoneApplicationPage` is shown as follows:

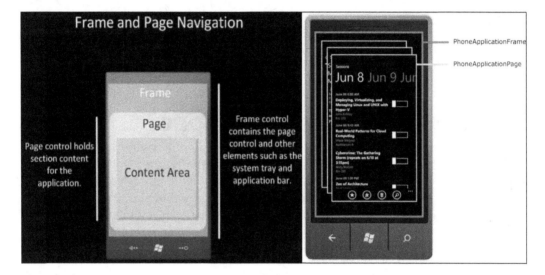

Frame and Page Navigation

PhoneApplicationFrame

This control is responsible for providing the Windows Phone look and feel to any
application. If we are developing a Windows Phone application, the application will
get only one `PhoneApplicationFrame`. Following are some of the characteristics of
the frame:

- It exposes certain properties of the page it hosts. For example, orientation
  of the screen.

- It exposes a client area where phone application pages are rendered.

- It exposes facilities for navigation between pages. For example, it exposes the Navigate() method, which can be used to move between pages.

- It reserves space for the status bar and the application bar.

Phone application page

A page fills the content region of a frame. It will contain contents or links to other pages. Applications can also contain pop ups, message boxes, or splash screens. A movement is considered as navigation only when it is between two pages. If there is a movement from the splash screen to a page, it is not considered as navigation, rather it's a transition.

Hub and Spoke navigation model

The navigation model available in Windows Phone is often referred to as a Hub and Spoke system of navigation. What this means is, as mentioned earlier, unless we as developers provide explicit links to other pages, users must use the hardware back button to navigate back to the page they viewed previously. Users always move forward through the pages. This type of model is very similar to how a web browser displays and navigates the web page history.

The navigation system in Windows Phone tracks each page a user visits and makes an entry in what is called as back stack. As the name goes, it's a stack of recently visited page history with the first entry being the latest visited page. When a user presses the hardware back button, the last saved page that is the first entry in the back stack is rendered. Back stack can have unlimited entries of visited pages and there is no limit to how many can be placed in the back stack.

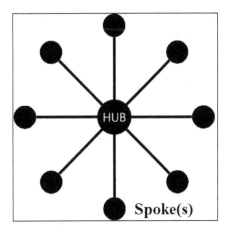

Hub and Spoke navigation model

For example, let us consider that we have three pages in our application – P1, P2, and P3. Consider the following navigation action:

- P1 to P2 that is P1 > P2
- P2 to P1 that is P2 > P1
- P1 to P3 that is P1 > P3
- P3 to P1 that is P3 > P1

The back stack created as part of this navigation is P1, P2, P1 and P3. The views in the back stack are just snapshots of the page at that point of navigation. So if you keep pressing the hardware back button, you will be navigated through all the pages in the back stack until we exit the application.

F# XAML Item Templates

In *Chapter 1*, *Setting up Windows Phone Development with F#*, we talked about certain prerequisites to be downloaded for developing Windows Phone applications using F#. They are:

- F# and C# Windows Phone App (Silverlight) Template:

    `http://gnl.me/FSharpWPAppTemplate`

- F# and C# Windows Phone List App (Silverlight) Template:

    `http://gnl.me/FSharpWPListAppTemplate`

- F# and C# Windows Phone Panorama App Template:

    `http://gnl.me/FSharpWPPanoramaAppTemplate`

- F# XAML Item Templates:

    `http://gnl.me/FSharpXAMLTemplate`

We have seen the first three templates in depth in *Chapter 2*, *F# Windows Phone Project Overview*. In this section, we will take a look at the fourth template namely *F# XAML Item Templates*. This is necessary to know, as we will need to do more than one page in our applications to understand the navigation system of Windows Phone.

You can download F# XAML Item Templates from the following URL: `http://gnl.me/FSharpXAMLTemplate`. If you have installed the template, open Visual Studio and go to **Tools** | **Extension Manager**. You should see the template in the **Installed Extensions** | **Templates** section.

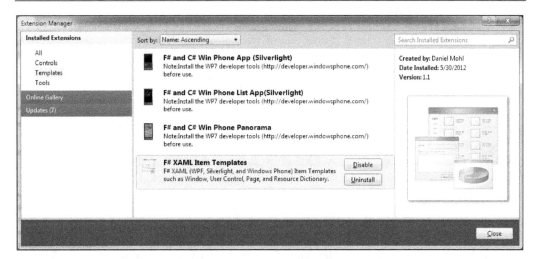

What this template does is provides the following XAML templates to be used in a Windows Phone F# application.

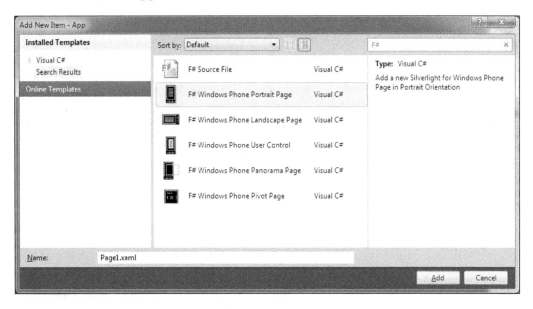

As you can see, this template provides different page templates, such as **Portrait**, **Landscape**, **Panorama**, and **Pivot** pages. So when we have to add a new page, just right-click on the project, select **Add | New Item...**, and pick the appropriate page template to work with.

Navigating between pages using a hyperlink button

One of the easiest ways to perform navigation in Windows Phone applications is to make use of a control called HyperlinkButton. If you recall, we studied a bit about HyperlinkButton in *Chapter 3, Working with Windows Phone Controls*. HyperlinkButton is a button that contains a hyperlink, which when clicked will navigate to the URI provided as part of the hyperlink. HyperlinkButton contains a property called NavigateUri, which is used to set the URI of the page we wish to navigate to. That's all that is there to perform the navigation. Let's see this in action now.

Let's create a new F# Windows Phone project. As usual, we will have MainPage. xaml and AppLogic.fs created in the project. Now, to perform navigation we will need one more page. The following steps will help you create a new page, navigate from MainPage to AnotherPage, and navigate from AnotherPage to MainPage:

1. Right-click on the **App** project and select **Add | New Item...**:

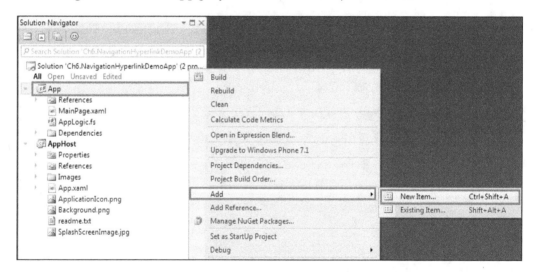

2. In the **Add New Item** window select **F# Windows Phone Portrait Page**:

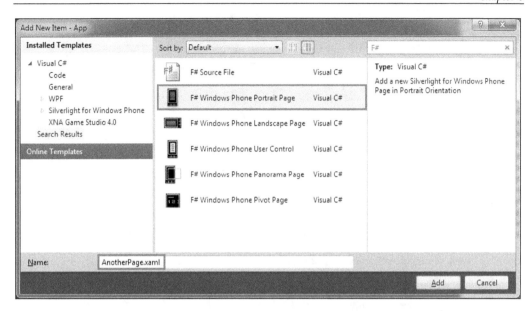

3. Now that we have **AnotherPage.xaml** created, we need a class file to handle the logic for it. Right-click on the **App** project and select **Add | New Item** from the menu. In the **Add New Item** dialog, select **F# Source File** and name the file as AnotherPageLogic.fs, as shown in the following screenshot:

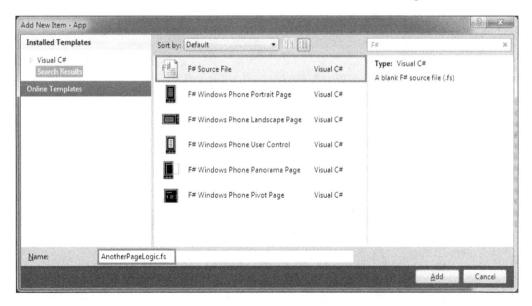

4. Open `AnotherPageLogic.fs` and add the following code:

```
namespace WindowsPhoneApp

open System
open System.Net
open System.Windows
open System.Windows.Controls
open System.Windows.Documents
open System.Windows.Ink
open System.Windows.Input
open System.Windows.Media
open System.Windows.Media.Animation
open System.Windows.Shapes
open System.Windows.Navigation
open Microsoft.Phone.Controls
open Microsoft.Phone.Shell

type AnotherPage() as this =
    inherit PhoneApplicationPage()

    // Load the Xaml for the page.
    do Application.LoadComponent(
        this,
        new System.Uri(
            "/WindowsPhoneApp;component/AnotherPage.xaml",
            System.UriKind.Relative
            )
        )
```

5. In `MainPage.xaml`, let's place a hyperlink button and give it a name. Here is the XAML snippet:

```
<HyperlinkButton  Content="Go To Another Page"
x:Name="MyHyperlink" />
```

6. In `AppLogic.fs`, find the type `MainPage` and add the following code:

```
let button:HyperlinkButton = this?MyHyperlink
    do button.NavigateUri <- new System.Uri(
        "/WindowsPhoneApp;component/AnotherPage.xaml",
        System.UriKind.Relative
    )
```

What we are doing here is, we are setting the `NavigateUri` property of `HyperlinkButton` to `AnotherPage.xaml`.

7. So on `MainPage.xaml`, we have a hyperlink, which when clicked will navigate to `AnotherPage.xaml`.

8. Open `AnotherPage.xaml` and place `HyperlinkButton`. Let's name the hyperlink button. The XAML snippet for it is shown as follows:

```
<HyperlinkButton  Content="Go To Main Page"
            x:Name="MyHyperlink" />
```

9. Open `AnotherPageLogic.fs` and add the following code snippet:

```
let button:HyperlinkButton = this?MyHyperlink
    do button.NavigateUri <- new System.Uri(
        "/WindowsPhoneApp;component/MainPage.xaml",
        System.UriKind.Relative
    )
```

10. We are setting the `NavigateUri` property of the hyperlink button to `MainPage.xaml`.

Here is what we have created so far. In `MainPage`, we have a hyperlink button with its `NavigateUri` set to `AnotherPage`. In `AnotherPage`, we have a hyperlink button with its `NavigateUri` set to `MainPage`. So at runtime, we will be able to navigate from `MainPage` to `AnotherPage` and from `AnotherPage` to `MainPage`. The screenshot of the app at runtime is shown as follows:

In `AnotherPage.xaml` there is no need to have an explicit navigation action to go back to `MainPage.xaml`. The hardware back button can be used to go back from `AnotherPage` to `MainPage`, and is the prescribed method to handle back navigation in applications.

Navigating between pages using NavigationService

In the previous section, we saw how to navigate using the hyperlink button, which is present on a page. As seen from the demo, we always set the target URI upfront as part of the `NavigateUri` property. But it is not always that we would need to use a hyperlink button for navigation. Instead, it's common for an application to have a navigation action on a button click or a tap on a list item. In such cases we cannot set the target URI upfront, rather we would handle those events and then navigate to the target page.

Fortunately, Windows Phone navigation system has a service named `NavigationService`, which helps us with the navigation. The `NavigationService` class provides methods, properties, and events to support navigation within an application. Two main methods of the `NavigationService` class are `Navigate` and `GoBack`. The `Navigate` method, as the name suggests, will navigate to a given URI, and `GoBack` will go back a screen that will bring into the forefront the latest page in the back stack.

We will consider the same example we did in the previous section. Instead of a hyperlink we will use a button to navigate. All the code and steps will be the same except that we will have a button in `MainPage` and a button in `AnotherPage`.

1. In `MainPage.xaml`, let's place a button and give a name to it. The XAML snippet for it is shown as follows:

```
<Button x:Name="NavigateButton"
        Content="Go To Another Page"
        Width="300" Height="100" />
```

2. In `AppLogic.fs`, find the `MainPage` type and handle the click event of the button. In the click event handler, we will use the `NavigationService.Navigate()` method to navigate to `AnotherPage.xaml`. The code snippet is shown as follows:

```
let button : Button = this?NavigateButton
    do button.Click.Add(fun _ ->
        this.NavigationService.Navigate(
            new System.Uri(
                "/WindowsPhoneApp;component/AnotherPage.xaml",
                System.UriKind.Relative
            )
        ) |> ignore
    )
```

3. In `AnotherPage.xaml`, let's place a button and give a name to it. The XAML snippet for it is shown as follows:

```
<Button x:Name="NavigateButton"
        Content="Go To Main Page"
        Width="300" Height="100"/>
```

4. In `AnotherPageLogic.fs`, handle the click event of the button. In the click event handler we will use the `NavigationService.GoBack()` method to navigate back to `MainPage.xaml`. The code snippet is shown as follows:

```
let button : Button = this?NavigateButton
    do button.Click.Add(fun _ ->
        if(this.NavigationService.CanGoBack) then
            do this.NavigationService.GoBack()
    )
```

We first check to see if there is any page available in the back stack or not. The `CanGoBack` property will let us know if there are any pages in the back stack. We then make use of the `GoBack()` method to go one page back in the navigation.

5. The screenshot at runtime is shown as follows:

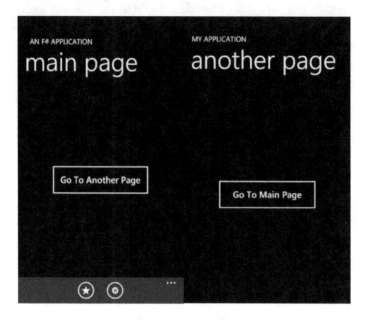

Passing data between pages

We saw how to carry out page-to-page navigation using two techniques in the previous sections. One of the requirements that we may come across in a day-to-day situation is to pass data from one page to another – the way we do it in web applications. For example, let's assume that we have a textbox on one page, and we want to pass the text typed in the textbox to another page and display it there. So while performing navigation, we would like to pass data too. We will look at how this is done in the Windows Phone navigation model.

The Windows Phone navigation model also supports the passing of data from one page to another. This is made possible by passing the data as a query string in the URI of the target page.

The Windows Phone navigation model provides two override methods in `PhoneApplicationPage`, which we can use to work with navigation actions. The two override methods are as follows:

Method	Description
OnNavigatedFrom()	Called whenever a user navigates away from a page
OnNavigatedTo()	Called whenever a user navigates to a page

We can pass data in the URI when we use `NavigationService.Navigate()`, and in the target page we can override the `OnNavigatedTo()` method to inspect the query string. In order to grab the query string, which may have been passed in the URI, we need to make use of a class called `NavigationContext`. `NavigationContext` helps us to retrieve the query string values for a navigation action. It exposes a property called `QueryString`, which is nothing but a collection of query string values in key/value pairs.

Let us now build a demo app, which will let us know how to pass data between pages. We will create a project with two pages (`MainPage.xaml` and `AnotherPage.xaml`) that are pretty similar to what we saw in the previous sections. One page will contain a textbox so that the user can type something, then we will navigate to one more page and display the text entered on the first page.

1. In `MainPage.xaml`, place a `Textbox` tag and a `Button` tag. Give a name to the `Textbox` and `Button` tag. The XAML snippet for `MainPage` is shown as follows:

```
<Grid x:Name="ContentGrid" Grid.Row="1">
        <StackPanel>
            <TextBox x:Name="txtMessage"
                    Height="80" Width="350" />
            <Button x:Name="btnNavigate"
                    Content="Go to Another Page"
                    Height="80" Width="350" />
        </StackPanel>
    </Grid>
```

2. In the button click event handler, we will pass the text typed in the textbox to `AnotherPage.xaml` as a query string in the URI. The code snippet for the button click event handler is shown as follows:

```
let textbox:TextBox = this?txtMessage
    let button:Button = this?btnNavigate
    let mutable targetUri:string =
        "/WindowsPhoneApp;component/AnotherPage.xaml"
    do button.Click.Add(fun _ ->
        let targetUri =  targetUri +
                            "?message=" + textbox.Text
        this.NavigationService.Navigate(
            new System.Uri(
                targetUri,
                UriKind.Relative
```

```
        )
    ) |> ignore
)
```

3. As you can see, we build the target URI and pass a query string named `message` with its value read from the textbox. We then navigate to `AnotherPage` with the `Navigate()` method.

4. Now let us see the XAML for `AnotherPage.xaml`. We will have a text block and the data passed from `MainPage` will be shown in this text block. The XAML snippet is shown as follows:

```
<Grid x:Name="ContentPanel"
            Grid.Row="1"
            Margin="12,0,12,0">
        <StackPanel>
            <TextBlock Text="Data Entered:"
                        HorizontalAlignment="Center"
                        Style="{StaticResource
PhoneTextTitle2Style}"/>
            <TextBlock x:Name="txtMessage"
                        HorizontalAlignment="Center"
                        Style="{StaticResource
PhoneTextTitle2Style}"/>
        </StackPanel>
    </Grid>
```

5. In the logic file of `AnotherPage.xaml`, what we do is we override the `OnNavigatedTo()` method and try to get the value of the data passed from the query string. The code snippet is shown as follows:

```
let textbox:TextBlock = this?txtMessage
    override this.OnNavigatedTo(e) =
        do base.OnNavigatedTo(e)
        let mutable message:string = ""
        do this.NavigationContext.QueryString.TryGetValue(
                "message",
                &message
            ) |> ignore
        do textbox.Text <- message
```

As you can see, all we are doing is getting the query string's value in the `OnNavigatedTo()` method and assigning it to the `TextBlock` class for it to be displayed.

6. The screenshot at runtime is shown as follows:

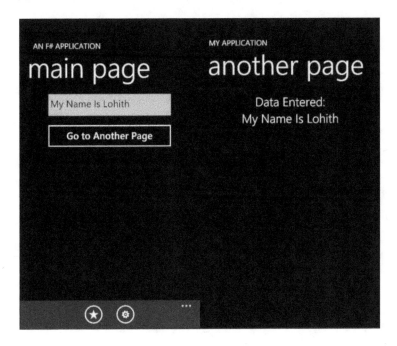

Summary

To wind up this chapter, let's go over what we learnt in this chapter. We took a look at one of the basic features of any application, that is, navigation, which means to move between pages of an application. We had a look at the Windows Phone navigation model. We learnt the different ways of performing navigation between the pages, that is, by using a hyperlink versus using the NavigationService class from code. Also, we looked at how to pass data from one page to another and how to read the data from a query string. I hope you now understand how navigation takes place in Windows Phone and how to pass and read data.

7
Windows Phone and Data Access

In this chapter, we will look at one of the basic needs of any application – data. We will learn how to store and retrieve data for use in an application. This chapter will help us understand the different techniques available to store and retrieve data and will also try to answer the question of determining which approach we should use depending on certain requirements.

Data sources

In this section, we will understand the different sources or origins of a particular piece of data. The data of interest for any application can reside at several locations. For example:

- Read-only data can be stored in a local file within the application
- Windows Phone provides an isolated storage where user-specific data can be stored locally on the phone
- Data can also be stored on the Internet, which can then be accessed using web services

The following figure depicts data source options available for a Windows Phone application:

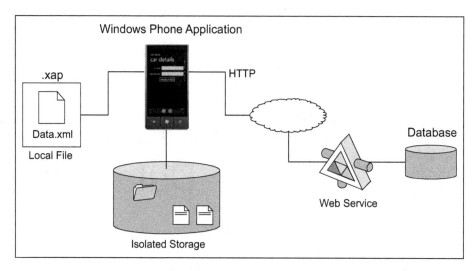

When we say local files, it means the file is packaged along with the application. Normally, we package a read-only data as a local file which is used by the application. Files such as text (.txt extension) or XML (.xml extension) are used for these purposes. Usually these local files are compiled either as a resource file or as a content file. We will look at each one of them in the following sections.

Resource files

Resource files are files which are embedded as resources in an application assembly. The advantage of following this mechanism is that the file will always be available to the application. But the start-up time of the application may increase when we use resource files. A file is designated as an embedded resource by setting its property **Build Action** to **Embed Resource** within the IDE. The embedded resource file can then be accessed by using a GetResourceStream method from the Application object, that is, Application.GetResourceStream.

Let's consider a demo to learn more about this mechanism. The following steps will help you create the demo:

1. Create a new **F# Windows Phone (Silverlight)** project. Give it a name.

2. In the **App** project, let's add an image as a new item. Right-click on the **App** project, go to **Add**, and then select **Exiting Item**. Select any image from your hard disk and add it to the project.

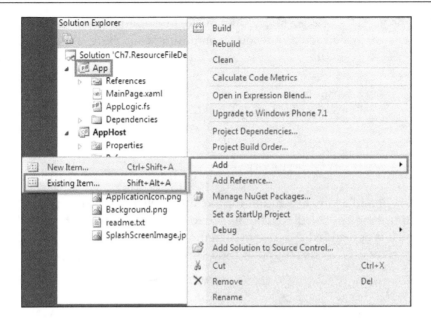

3. Right-click on the newly added item and select **Properties**. In the **Properties** window, make sure that the **Build Action** item is set to **Resource**. Leave the rest of the items as is.

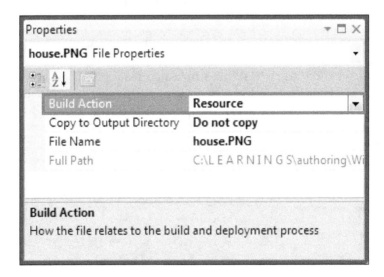

4. In the `MainPage.xaml` file, add an image and a button. The idea is to load the image we just added as a resource on button click. Here is the XAML snippet of `MainPage.xaml`:

```
<Grid x:Name="ContentGrid" Grid.Row="1">
    <StackPanel>
      <Button x:Name="btnImageLoader"
              Content="Load Image" />
      <Image x:Name="img" />
    </StackPanel>
</Grid>
```

5. We will add an event handler for the button click. In the event handler, we will use the `Application.GetResourceStream()` method to read the image which is embedded as a resource in the **App** project and load it in the `Image` tag. Here is the code snippet for the same:

```
let button : Button = this?btnImageLoader
    let image : Image = this?img
    let resourceUri = "WindowsPhoneApp;component/house.png"
    do button.Click.Add(fun _ ->
        let uri = new Uri(resourceUri, UriKind.Relative)
        let streamInfo = Application.GetResourceStream(uri)
        let bitmapImage = new BitmapImage()
        bitmapImage.SetSource(streamInfo.Stream)
        do image.Source <- bitmapImage
    )
```

As you can see, we build the URI of the embedded resource. This is called the pack URI and it follows a convention like `"<namespace>;component/<resource name with extension>`. Then we use `Application.GetResourceStream` to get the resource stream. Since in this example we know that the resource is an image, we create an instance of `BitmapImage` and set its source to the stream obtained from the resource. Finally, we set the source of the image we have on the XAML to the bitmap image we just created.

Here is the screenshot of the application at runtime:

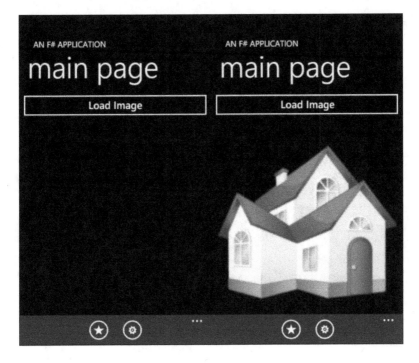

Recommended scenarios for using resource files are:

- When the application start-up time is not a concern
- When there is no need to update the resource being embedded into the assembly after it is compiled into an assembly
- If you do not want to increase the number of physical file dependencies during deployment

So in this section, we saw how to embed a resource file and at runtime retrieve the same and use it. In the next section, we will see how to embed a resource as content and use the same.

Content files

Content file is any file which is not embedded into a project assembly or a library assembly instead it is packaged as is with the application and after deployment can be found at the application root.

To specify a file as a content file, we need to go to the file's property and change the **Build Action** to **Content**. This tells the compiler to not package it as a resource, but rather to keep the physical file as is in the deployment package. The content file can be accessed relative to the application root once installed on the device.

Let's create a demo to understand how to package a file as content. We will create a simple text file, let's say "Terms of Use" and add it to the project as content. We will then read this text file at runtime and show it in a text block. The following steps will help you create the demo app:

1. Create an **F# Windows Phone Application** project and give it a name.

2. In the **App** project, right-click on the project and go to **Add | New Item**. In the **Add New Item** dialog, select the **Text File** item template and give it a name, let's say, Terms.txt.

3. Right-click on the `Terms.txt` file we added in the previous step. Select **Properties** from the context menu. In the **Properties** dialog box, make sure that **Build Action** is set to **Content** and **Copy to Output Directory** is set to **Copy always**.

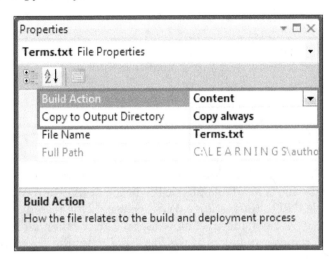

Enter some text in the text file. Don't worry about the specifics here. This is just a demo file without any real-world content.

4. Now in the `MainPage.xaml` file let us add a button and a text block. The XAML snippet is given as follows:

```xml
<Grid x:Name="ContentGrid" Grid.Row="1">
  <StackPanel>
    <Button x:Name="btnTermsLoader"
            Content="Show Terms of Use" />
    <Border BorderThickness="1"
            BorderBrush="LightGray"
            Margin="12">
      <TextBlock x:Name="txtTerms"
                 TextWrapping="Wrap"
                 HorizontalAlignment="Stretch"/>
    </Border>
  </StackPanel>
</Grid>
```

5. Next we handle the click event of the button by adding an event handler. In the event handler, we will try to open the content file as a text file, read the content, and show it in the text block. Here is the code snippet:

```
let button : Button = this?btnTermsLoader
    let textBlock : TextBlock = this?txtTerms
    do button.Click.Add(fun _ ->
        let resourceUri = new Uri("Terms.txt", UriKind.Relative)
        let resourceStream = Application.
GetResourceStream(resourceUri)
        let streamReader = new StreamReader(resourceStream.Stream)
        let terms = streamReader.ReadToEnd()
        do textBlock.Text <- terms
    )
```

First we build the resource URI. Notice that we don't have to use the pack URI in this case because the file is available physically in the application root, so it's enough we give the relative URI to the resource. We then access the resource as a stream, instantiate a `StreamReader` (since we know that we are dealing with a text file in this example), and read the contents of the file. Finally, set the `Text` property of the `TextBlock` to the file content.

Here is a screenshot of the app at runtime:

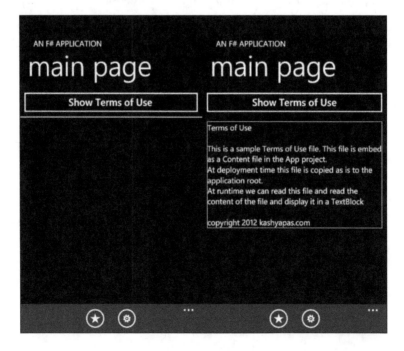

Recommended scenarios for using content files are:

- If we have a need to update the content file without recompiling the assembly
- If we need the application to start up quickly

Isolated storage

Often, we would need to store and retrieve user-specific information or data as part of the application. Windows Phone operating system provides a mechanism to deal with user-level data in an isolated environment called **isolated storage**.

As the name indicates, this is a storage which is isolated from other applications running on the device. Each application gets its own isolated storage where user-specific data or information can be stored and retrieved. Applications do not have access to an operating system's filesystem but have access to the isolated storage. This is like a virtual filesystem that can be accessed only by the application associated with it.

There are two ways in which isolated storage can be used. They are:

- **IsolatedStorageSettings** – saving/retrieving data as key/value pairs. Very similar to the `AppSettings` mechanism in most of the .NET application.
- **IsolatedStorageFile** – saving/retrieving data as files.

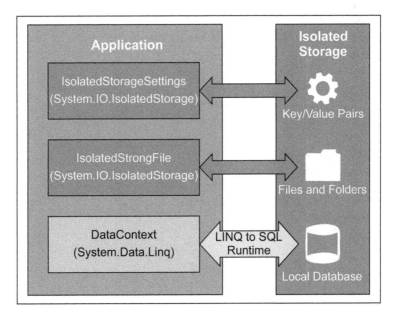

IsolatedStorageSettings

Whenever we have a scenario to store data in a key/value pair manner, we can make use of the `IsolatedStorageSettings` class. `IsolatedStorageSettings` is nothing but a dictionary which holds data in key and value pairs. This type of storage settings is ideal for storing small pieces of data, for example, application settings, which may need to be accessed when the application loads and exits.

The `IsolatedStorageSettings` class exposes many properties and methods. But the important methods which we need to know about are as follows:

- **Add**: Used to add an entry to the dictionary with a key and value pair
- **Contains**: Used to check if a key is present in the dictionary
- **Remove**: Used to remove a specified key from the dictionary

Let us create a demo app which showcases the use of `IsolatedStorageSettings`. The app will add, retrieve, and remove a key/value pair from the isolated storage. We will have a textbox for the user to enter a value. We will store that with a key; the value being the user-entered text. The user can modify the value and can delete it from the settings. The following steps will let you create the app:

1. Create a new **F# Windows Phone Application** project. Give it a name.
2. Open `MainPage.xaml` and add one textbox and three buttons. Add one text block to show the stored value in the settings. The XAML snippet for it is shown as follows:

```
<Grid x:Name="ContentGrid" Grid.Row="1">
  <StackPanel>
    <TextBox x:Name="txtData" />
    <Button x:Name="btnSave"
            Content="Save Data" />
    <Button x:Name="btnDisplay"
            Content="Display Data" />
    <Button x:Name="btnRemove"
            Content="Remove Data" />
    <TextBlock Margin="12"
               x:Name="txtDisplayData" />
  </StackPanel>
</Grid>
```

3. In the `AppLogic.fs` file, let's add code to handle the Save, Display, and Remove button's click events. In the Save button's click event handler, we just check if the key is present in the dictionary or not. If already present, we update the value; else we create a new key and add the value. In the Display click event handler, we just read the value from the dictionary and show it in a text block. In the Remove click event handler, we just remove the key from the dictionary. Here is the code snippet:

```
let settings : IsolatedStorageSettings =
        IsolatedStorageSettings.ApplicationSettings
    let data : TextBox = this?txtData
    let displayBlock : TextBlock = this?txtDisplayData
    let buttonSave : Button = this?btnSave
    let buttonDisplay : Button = this?btnDisplay
    let buttonRemove : Button = this?btnRemove

    do buttonSave.Click.Add(fun _ ->
        if(settings.Contains("UserData")) then
            settings.["UserData"] <- data.Text
        else
            settings.Add("UserData", data.Text)
        settings.Save()
    )

    do buttonDisplay.Click.Add(fun _ ->
        if(settings.Contains("UserData")) then
            let settingsValue = settings.["UserData"] :?> String
            do displayBlock.Text <-
                String.concat " " ["USER DATA:"; settingsValue]
    )
    do buttonRemove.Click.Add(fun _ ->
        if(settings.Contains("UserData")) then
            settings.Remove("UserData") |> ignore
            displayBlock.Text <- ""
            data.Text <- ""
    )
```

The app at runtime can be seen in the following screenshot:

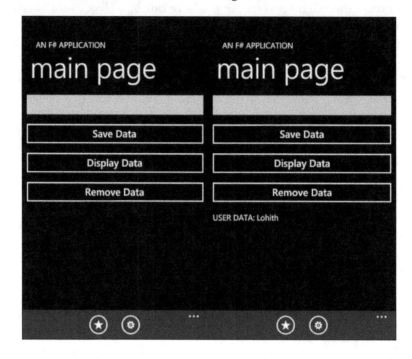

IsolatedStorageFile

In the previous section, we saw how to save simple data which is user-specific using `IsolatedStorageSettings`. But we do not always have a scenario where we have to deal with simple key/value pairs. Consider a scenario where we would like to save an image file required for the application we are developing. As stated previously, the Windows Phone applications cannot access the operating system's filesystem. So the Windows Phone OS provides an isolated storage where we can store and retrieve files. The class that helps us achieve this is known as `IsolatedStorageFile`. `IsolatedStorageFile` represents a storage area which can contain files and folders and which is accessible only by the application that owns it.

The `IsolatedStorageFile` class along with `IsolatedStorageFileStream` makes it possible to work with files and directories in the isolated storage. `IsolatedStorageFile` is used to access the file and directories whereas `IsolatedStorageFileStream` is used to read the content of the file. The `IsolatedStorageFile` class exposes many properties and methods, but we typically use the following methods that are of the utmost importance to us now. They are:

Method	Description
GetUserStore ForApplication	Obtains user-scoped isolated storage for use by an application
FileExists	Determines whether the specified path refers to an existing file in the isolated store
CreateFile	Creates a file in the isolated store
OpenFile	Opens a file at a specified path. Returns an IsolatedStorageFileStream object that contains the file's stream
DeleteFile	Deletes a file in the isolated store
DirectoryExists	Determines whether the specified path refers to an existing directory in the isolated store
CreateDirectory	Creates a directory in the isolated storage scope
DeleteDirectory	Deletes a directory in the isolated storage scope
Remove	Removes the isolated storage scope and all its contents

Both the IsolatedStorageFile and IsolatedStorageFileStream instances, if created in the code, should be disposed of after their use. This is necessary as these are IO-based classes and we don't want to hang on to the file and file stream for a long time. We can use the using statement of the language to achieve this. What using does is, as soon as the object has been done with the usage, dispose will be called automatically.

Let us create a sample application to demonstrate the usage of IsolatedStorageFile and IsolatedStorageFileStream. In this demo app, we will have a textbox in the main page where users can enter a string of text, write it to a file, and then read the contents of the text file. Follow the following steps to create the demo app:

1. Create a new **F# Windows Phone Application** project. Give it a name.

2. In the MainPage.xaml file, add one textbox for the user to enter text and one button to write the entered text to the file. Add one text block to show the contents of the file and one button to initiate the read action. The XAML snippet for this is shown as follows:

```
<Grid x:Name="ContentGrid" Grid.Row="1">
  <StackPanel>
    <TextBox x:Name="txtData" />
      <Button x:Name="btnWrite"
         Content="Write" />
        <Button x:Name="btnRead"
```

```
        Content="Read"
        Margin="0,50,0,0" />
    <TextBlock x:Name="txtReadBlock"
        Margin="12"/>
    </StackPanel>
</Grid>
```

3. In the `AppLogic.fs` file under the type `MainPage`, let's add an event handler for the `btnWrite` button click event. In this event, we will first get hold of the store, create a directory, create the file, and write the text entered by the user. The code snippet for it is given as follows:

```
let data : TextBox = this?txtData
    let writeButton : Button = this?btnWrite
    do writeButton.Click.Add(fun _ ->
        let isoStore = IsolatedStorageFile.
GetUserStoreForApplication()
        if(isoStore.DirectoryExists("DemoFolder") = false) then
            isoStore.CreateDirectory("DemoFolder")
        use isoFileStream = new IsolatedStorageFileStream("DemoFol
der\\demofile.txt",
                                        FileMode.OpenOrCreate,
                                        isoStore)
        use writer = new StreamWriter(isoFileStream)
        writer.WriteLine(data.Text)
    )
```

4. Next, let's add an event handler for the Read button click event. In the event handler, we first get hold of the isolated store, then get the stream to the file, and finally read the content from the stream using a stream reader and display it in the text block. The code snippet is given as follows:

```
let readBlock : TextBlock = this?txtReadBlock
    let readButton : Button = this?btnRead
    do readButton.Click.Add(fun _ ->
        let isoStore = IsolatedStorageFile.
GetUserStoreForApplication()
        use isoFileStream = new IsolatedStorageFileStream("DemoFol
der\\demofile.txt",
                                        FileMode.OpenOrCreate,
                                        isoStore)
        use reader = new StreamReader(isoFileStream)
        readBlock.Text <- reader.ReadToEnd()
    )
```

The app at runtime can be seen in the following screenshot:

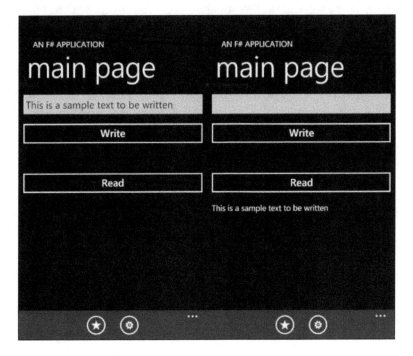

HTTP classes

In the beginning of this chapter, we talked about data and different data sources that we have. One of the data source or origins of data was the Internet. Assume that some server is hosting data and we need to consume that data from within our application. The data on the server will be exposed as a web service. In order for us to access the web services from Windows Phone applications, we can make use of the following classes provided by the .NET framework:

- `HttpWebRequest`
- `HttpWebResponse`
- `WebClient`

These classes can be found in the `System.Net` namespace. They provide us with the functionality required to send requests to any web service over the HTTP protocol. These classes should be used when we are consuming a web service hosted by a third party and the service response is XML or **JavaScript Object Notation (JSON)**. These classes can be used to build the request which match the format expected by the service.

Let's build a demo app which showcases the use of `WebClient`, a simple class when compared to `HttpWebRequest`/`HttpWebResponse`. We will connect to Flickr web service, do a photo search based on the user-entered text, retrieve one photo, and show that image on the page. The following steps will help you to create the demo app:

1. Create a new **F# Windows Phone Application** project. Give it a name.

2. In the `MainPage.xaml` file, add a textbox for the user to enter a search text and a button to search the Flickr photo service. The XAML snippet for it is given as follows:

```
<Grid x:Name="ContentGrid" Grid.Row="1">
  <StackPanel>
    <StackPanel Orientation="Horizontal">
      <TextBox x:Name="txtQuery"
               Height="74"
               Width="286" />
      <Button x:Name="btnSearch"
              Content="Search Flickr"
              HorizontalAlignment="Right"/>
    </StackPanel>
    <Image Margin="12"
           x:Name="img" />
  </StackPanel>
</Grid>
```

3. In the **App** project, add a reference to `System.Xml.Linq.dll`. Right-click on the project and select **Add Reference**. In the **Add Reference** dialog, search for **System.Xml.Linq** and click on the **Add** button.

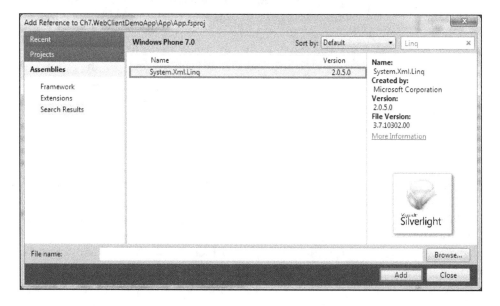

4. In `AppLogic.fs`, add the following namespace in the code-behind file:

```
open System.IO
open System.Xml.Linq
open System.Windows.Media.Imaging
```

Under the type `MainPage`, handle the click event of the search button. Add an event handler where we will first make an HTTP call, using the `WebClient` class, to the Flickr search API. Then get the response, parse the response to get the photo details, and update the source of the image tag. The code snippet for it is given as follows:

```
let query : TextBox = this?txtQuery
    let searchButton : Button = this?btnSearch
    let image : Image = this?img
    do searchButton.Click.Add(fun _ ->
        let searchUri = String.Format(baseURI, query.Text)
        let webClient = new WebClient()
        webClient.OpenReadCompleted.Add(fun args ->
            let stream : Stream = args.Result
            let xd = XDocument.Load(stream)
            let photoElement = xd.Element(GetXName "rsp")
                                 .Element(GetXName "photos")
                                 .Element(GetXName "photo")
            let imageUri = String.Format(baseImageUri,
                                         GetAttribute photoElement
"farm",
                                         GetAttribute photoElement
"server",
                                         GetAttribute photoElement
"id",
                                         GetAttribute photoElement
"secret")
            image.Source <- new BitmapImage(new Uri(imageUri))
        )
        do webClient.OpenReadAsync(new Uri(searchUri))
    )
```

As you can see, we create a new instance of the `WebClient` class. We then add an event handler for the `OpenReadCompleted` event. We get the result which is a stream object and XML format and use that to load it into an XDocument. We then parse the XML to reach the photo node. We then extract attributes from the photo node in order to build the image URI. Next we create a `BitmapImage` and provide it to the source of the image on the XAML.

We make a request to the Flickr search API using the `OpenReadAsync` method of the `WebClient` class by passing the search URI. The Flickr search API's URI is built as follows:

```
let baseURI = "http://api.flickr.com/services/rest/?"
            + "method=flickr.photos.search"
            + "&api_key=<your API key>"
            + "&per_page=1&page=1"
            + "&format=rest"
            + "&text={0}"
```

The search string entered by a user is passed as a text key in the query string. The screenshot of the app at the runtime is given as follows:

Summary

In this chapter, we saw different sources of data available for any application. We also saw the techniques required to store or read data under those different sources. Here is a suggestion that can be kept in mind in deciding the approach to follow for your Windows Phone application:

Scenario	Suggested approach
Embedding read-only data in the assembly	Use the local resource file approach
Updating read-only data without recompiling assemblies	Use content file approach
Store and retrieve user-specific information	Use isolated storage
Access data from a third-party REST service	Use HTTP classes
Access RSS feeds	Use HTTP classes

8

Launchers and Choosers

In all the applications we have done till now, we never used any of the built-in applications on the device. The built-in applications include apps such as the phone dialer or the data libraries, that is, the photo library. In this chapter, you will see what we mean by Launchers and Choosers. The Launchers and Choosers will help us to use the built-in apps or call the built-in apps right from our own Windows Phone applications.

Overview of Launchers and Choosers

Launchers and Choosers enable the Windows Phone applications to access the built-in applications and data stores on the device. For example, if we want to make a phone call from within our application, we would use what we call a Launcher; or if we want to access the data libraries on the device, such as the photo library, we would use what we call a Chooser.

The main difference between a Launcher and a Chooser is that Launchers don't return any value, whereas Choosers return a value. For example, we have a Launcher known as `EmailComposerTask` that starts the e-mail application and when it exits, the control is returned to the application that called it. We also have a Chooser called `CameraCaptureTask` that starts the camera application. After the user takes a picture, the camera application exits and returns the value of the photo that was taken.

When we call a Launcher or a Chooser, our Windows Phone application goes to the background, and the built-in application that the Launcher or Chooser launches will run in the foreground. When the launched built-in application exits, our application comes to the front. The process of sending an application to the background is technically known as deactivation, and the process of bringing an application to the foreground is known as activation.

The Launchers or Choosers are all in the `Microsoft.Phone.Tasks` namespace. So we will need to include this namespace in our code-behind file when we want to use the Launchers and Choosers.

Launchers

A **Launcher** is an API that launches one of the built-in applications, for example, a `WebBrowserTask` Launcher launches a web browser application. A user uses the Launcher's built-in application to finish his task. When the built-in application is launched, the user has the option of completing or cancelling the task. The following are the general steps for launching a Launcher:

- Create an instance of the Launcher task type.
- Set any required and optional properties of the task object. Then determine the behavior of the Launcher when invoked.
- Call the `Show` method on the task to launch the Launcher.

A list of Launchers available for the Windows Phone is given as follows:

Launcher	Description
BingMapsTask	Launches the Bing Maps application
BingMapsDirectionsTask	Launches the Bing Maps application and displays the driving directions between two points
EmailComposeTask	Enables users to send e-mail from an application
MarketplaceDetailTask	Launches the Windows Phone Marketplace client application and displays the details page for a specified application
MarketplaceHubTask	Launches the Windows Phone Marketplace client application with a hub specified, that is, either an application or the music hub
MarketplaceReviewTask	Launches the Windows Phone Marketplace client application and displays the review page for the current application
MarketplaceSearchTask	Launches the Windows Phone Marketplace client application and allows users to search the marketplace
MediaPlayerLauncher	Launches the media player and plays the specified media file
PhoneCallTask	Enables users to make a phone call from an application
SearchTask	Launches the web search application

Launcher	Description
ShareLinkTask	Enables users to share a link on a social network of their choice
ShareStatusTask	Enables users to share a status on a social network of their choice
SMSComposeTask	Enables users to send SMS from an application
WebBrowserTask	Launches the web browser application

BingMapsTask

This Launcher task, as the name goes, shows a Bing Maps application when launched. We can set the following properties on the task:

- `SearchTerm` – a string that will be used to perform a search on the map. This can also be an address.
- `Center` – an optional property to specify a coordinate to which the map should be centered. If not specified, the user's current location is used as the center.
- `ZoomLevel` – an optional property to specify the level of zoom when the map appears. It takes a value from 1 to 19.

A code snippet to use this task is shown as follows:

```
let task = new BingMapsTask()
task.SearchTerm <- "coffee"
//Omit the center property to use user's current location
//task.Center = new GeoCoordinate(47.6204, -122.3493)
task.ZoomLevel <- 2
task.Show()
```

BingMapsDirectionsTask

This Launcher task, when used, will show a Bing Maps application. As it is a maps application, we can use it to get driving directions from point A to point B. The map needs a start and end point; if we omit the start point, it will automatically use the current location as the start point. In order to set a start or end point, we need to provide the location as a string label and/or provide the geographic coordinates of the points. If no geographic coordinate points are specified, the map will use the label string of the location. A code snippet to use this task is shown as follows:

```
let task = new BingMapsDirectionsTask()
task.End = new LabeledMapLocation("Space Needle", null)
task.Show()
```

EmailComposeTask

This Launcher, as the name suggests, launches the Send Email application when launched. This task launches the e-mail application and displays a new e-mail message. The following properties are exposed by this task:

- `Subject` – an optional property which , when set, forms the subject of the new e-mail.

- `Body` – an optional property which, when set, forms the body of the new e-mail. At this moment only plain text is possible.

- `To` – an optional property which, when set, forms the recipient of the new e-mail.

- `Cc` – an optional property which, when set, forms the carbon copy recipients of the new e-mail.

The e-mail will be sent only when the user clicks on the send button in the e-mail application. A code snippet to use this task is given as follows:

```
let task = new EmailComposeTask()
task.Subject <- "message subject"
task.Body <- "message body"
task.To <- "recipient@example.com"
task.Cc <- "cc@example.com"
task.Show()
```

MarketplaceDetailsTask

This Launcher can be used to open the Marketplace client application from your own application. When launched, this task will show the details page of an application in the Windows Phone Marketplace. The following properties can be set on this task:

- `ContentIdentifier` – an optional property. If set, the value will be the GUID of any application.

- `ContentType` – an optional property. Enum of type `MarketplaceContentType` and contains two values: `Applications` and `Music`.

Since both the properties are optional, if they are launched without any properties set, this will show the current application's details page on the Marketplace. A code snippet to use this task is shown as follows:

```
let task = new MarketplaceDetailTask()
task.ContentIdentifier <- "c14e93aa-27d7-df11-a844-00237de2db9e"
task.ContentType <- MarketplaceContentType.Applications
task.Show()
```

MarketplaceHubTask

This Launcher can be used to open the Marketplace client application from your own application. Windows Phone Marketplace has two main content sections, namely, `Applications` and `Music`. So you can directly navigate to the Marketplace hub and launch either applications or music content from your own application. This task has one property that needs to be set, namely, `ContentType`. This is an enum of type `MarketplaceContentType` and contains the `Applications` and `Music` values. A code snippet to use this task is shown as follows:

```
let task = new MarketplaceHubTask()
task.ContentType <-  MarketplaceContentType.Music
task.Show()
```

MarketplaceReviewTask

This Launcher, as the name goes, is used to launch the current application's review page on the Windows Phone Marketplace. This will launch the Marketplace client application, navigate to current application details, and focus the review page. This does not have any properties. We just instantiate the task and call the `Show` method of the task. A code snippet to use this task is shown as follows:

```
let task = new MarketplaceReviewTask()
task.Show()
```

MarketplaceSearchTask

This Launcher is used to launch the Marketplace client application and show the search page to search the items in the Windows Phone Marketplace to see whether a search term is specified or not. We can set the content to search for. The content types that are supported for searching are `Applications` and `Music`. A code snippet to use this task is shown as follows:

```
let task = new MarketplaceSearchTask()
task.SearchTerms <- "bill analyzers"
task.ContentType <- MarketplaceContentType.Applications
task.Show()
```

MediaPlayerLauncher

As the name suggests, this Launcher is used to launch the media player on the device and play the media file that we specify. This allows us to enable playing music or video directly from our own application. We have to specify the name and the location of the media file on the device—isolated storage or within the application. We can also optionally specify the media player controls, such as rewind and stop. The media player controls can be specified as a bitwise combination of controls. The following properties can be set on the Launcher:

- `Media` – used to set the name of the media.

- `Location` – used to set the location of the media. This is a enum of type `MediaLocationType` that contains the `Data` and `Install` values. If the media is located in `IsolatedStorage`, the `Location` should be set to `Data`; or else, if the media file is part of the application, `Location` should be set to `Install`.

- `Controls` – used to set the media player controls as a bitwise combination.

A code snippet to use this task is shown as follows:

```
let mediaPlayerLauncher = new MediaPlayerLauncher()
mediaPlayerLauncher.Media <- new Uri("test.mp3", UriKind.Relative)
mediaPlayerLauncher.Location <- MediaLocationType.Install
mediaPlayerLauncher.Controls <- MediaPlaybackControls.Pause |||
MediaPlaybackControls.FastForward |||
MediaPlaybackControls.Rewind |||
MediaPlaybackControls.Skip |||
MediaPlaybackControls.Stop
mediaPlayerLauncher.Show()
```

PhoneCallTask

This Launcher task enables users to make a phone call from your application. We will need to set the name and the number of the person to call as part of the task properties. When the task is launched, it will display the phone dialer application on the device and also display the name and the number set. The call will not be placed until the user clicks on the call button. A code snippet to use this task is shown as follows:

```
let task = new PhoneCallTask()
task.DisplayName <- "Fred"
task.PhoneNumber <- "(555)-555-5555"
do task.Show()
```

SmsComposeTask

This Launcher is used to enable the users to send SMS messages from an application. When this task is invoked, it will launch the messaging application with the new SMS message displayed. We can optionally set the body of the message and the recipients of the message. The message will not be sent until the user clicks on the send button. A code snippet to use this task is shown as follows:

```
let task = new SmsComposeTask()
task.Body <- "this is a sample text message"
task.To <- "919999999999"
task.Show()
```

WebBrowserTask

This Launcher allows us to launch the web browser application on the device from an application and will display the URL we specify. This Launcher task provides the URL property to set the address of the website that we need to navigate to. A code snippet to use this task is shown as follows:

```
let task = new WebBrowserTask()
task.URL <- "http://windowsphone.com"
task.Show()
```

Choosers

In the previous section we took a look at Launchers and understood the different Launchers available in Windows Phone. In this section we will understand the counterpart of Launchers, which is known as **Choosers**.

Choosers are APIs that launch one of the built-in applications, such as Contacts or Camera from your application, allows the user to perform a task, and then returns the value selected by the user to your application. Like the Launchers, Choosers also provide the ability to complete or cancel the task for the user. When a Chooser is launched, the current application is sent back or deactivated and the launched built-in application takes focus. When the user finishes his task in the built-in application, it is closed and the calling application is brought back to the front, or is activated. All Choosers follow these general steps:

1. Create a new instance of the Chooser task type.

2. Add the Callback method to handle the completed event.

3. Set the required or optional properties of the task.
4. Call the `Show()` method on the task.
5. Handle the data returned by the `Callback` method.

Chooser	Description
AddressChooserTask	Launches the Contacts application and is used to obtain the physical address of a contact
CameraCaptureTask	Launches the Camera application
EmailAddressChooserTask	Launches the Contacts application and is used to obtain the e-mail address of a contact
PhoneNumberChooserTask	Launches the Contacts application and is used to obtain the phone number of a contact
SaveContactTask	Launches the Contacts application and is used to save a contact
SaveEmailAddressTask	Launches the Contacts application and is used to save an e-mail address to a new or existing contact
SavePhoneNumberTask	Launches the Contacts application and is used to save a phone number to a new or existing contact
SaveRingtoneTask	Launches the Ringtone application and is used to save a ringtone to the system ringtones list

AddressChooserTask

`AddressChooserTask` is a task that allows an application to obtain the address of a contact selected by a user. When invoked, it launches the Windows Phone's built-in Contacts application, allows the user to select a Contact, and when done, it will return the physical address of the contact back to the application. As with all the Chooser tasks, this also has a completed event in which the selected address is sent as a result. The code snippet for this Chooser is given as follows:

```
let task = new AddressChooserTask()
task.Completed.Add(fun result ->
  if(result.TaskResult = TaskReuslt.OK) then
         textBlock.Text <- result.Address
)
task.Show()
```

CameraCaptureTask

As the name goes, this Chooser task allows the users to take a photo from your application by launching the built-in Camera application. When the user completes the task, that is, finishes clicking a photo, the clicked photo is sent as a result in the `Completed` event. The code snippet for this Chooser task is given as follows:

```
let task = new CameraCaptureTask()
task.Completed.Add(fun result ->
  if(result.TaskResult = TaskResult.OK) then
  let bmp = new BitmapImage()
      bmp.SetSource(result.ChosenPhoto)
)
task.Show()
```

EmailAddressChooserTask

This Chooser task allows an application to obtain an e-mail address of a contact selected by a user from the phone's contacts list. This launches the built-in Contacts application. When invoked, a user can select a contact and the `Completed` event will return the e-mail address as the result. The code snippet for this Chooser task is shown as follows:

```
let task = new EmailAddressChooserTask()
task.Completed.Add(fun result ->
textBlock.Text <- result.Email
)
task.Show()
```

PhoneNumberChooserTask

This Chooser task allows an application to obtain the phone number of a contact selected by a user from the phone's contacts list. This launches the built-in Contacts application. When invoked, a user can select a contact and the `Completed` event will return the phone number in the result. Here is the code snippet for this Chooser task:

```
let task = new PhoneNumberChooserTask()
task.Completed.Add(fun result ->
textBlock.Text <- result.PhoneNumber
)
task.Show()
```

SaveEmailAddressTask

We can use this Chooser task to enable a user to save an e-mail address from your application. When invoked, this task launches the built-in Contacts application. The code snippet for this Chooser task is shown as follows:

```
let task = new SaveEmailAddressTask()
task.Email <- "someone@abc.com"
task.Completed.Add(fun result ->
  if(result.TaskResult = TaskResult.OK) then
    MessageBox.Show("Saved.")
)
task.Show()
```

Summary

In this chapter, we looked at one of the most interesting features supported as part of the Windows Phone operating system for custom application development. Launchers and Choosers bridge the gap between the custom applications and the built-in applications by allowing the custom applications to invoke the built-in applications by way of using the Launcher and Chooser tasks. While Launchers don't return any value when a user finishes the task, the Choosers, on the other hand, return the value selected by the user at the end of a task. By using the Launchers and Choosers in an application, we provide a consistent user experience through the Windows Phone platform.

9
Windows Phone Sensors

In this chapter, we will look at one of the coolest features that the Windows Phone platform has to offer, namely, sensors. Windows Phone supports multiple sensors and these sensors allow apps to determine the orientation and motion of the device. With sensors, it is possible to develop apps that make the physical device itself an input. Typical uses of these sensors are in games, location-aware apps, and so on. The Windows Phone platform provides APIs to retrieve data from the individual sensors. We will have a brief overview of the following sensors in this chapter:

- Accelerometer
- GPS

Accelerometer

An **accelerometer** is a sensor that can measure the intensity and direction of the acceleration force that the phone experiences. The intensity reading is provided as a decimal value ranging from -1.0 to 1.0 for the X, Y, and Z axes on the phone. This sensor monitors the acceleration force experienced width-wise, length-wise, and depth-wise. In order to detect the direction of force, the X, Y, and Z values must be compared to one another.

Let's create a demo app to understand how to use the accelerometer sensor:

1. Create a new project of the type **F# Windows Phone Application (Silverlight)** and give it a name.
2. Right-click on the **App** project and add a reference to **Microsoft.Devices. Sensors** from the list.

3. In `MainPage.xaml`, add three text blocks to display the X, Y, and Z reading. The XAML snippet for the same is shown as follows:

```
<TextBlock  Height="30" HorizontalAlignment="Left"
        Margin="20,100,0,0" Name="xTextBlock"
            Text="X: 1.0" VerticalAlignment="Top"
        Foreground="Red" FontSize="28" FontWeight="Bold"/>
<TextBlock  Height="30" HorizontalAlignment="Center"
            Margin="0,100,0,0" Name="yTextBlock"
            Text="Y: 1.0" VerticalAlignment="Top"
            Foreground="Green" FontSize="28" FontWeight="Bold"/>
<TextBlock  Height="30" HorizontalAlignment="Right"
            Margin="0,100,20,0" Name="zTextBlock"
            Text="Z: 1.0" VerticalAlignment="Top"
            Foreground="Blue" FontSize="28" FontWeight="Bold"/>
```

4. In `AppLogic.fs`, add the `Microsoft.Devices.Sensors` namespace shown as follows:

```
open Microsoft.Devices.Sensors
```

5. Create an instance of the `Accelerometer` class:

```
let accelerometer : Accelerometer
 = new Accelerometer()
```

6. The accelerometer exposes an event called `ReadingChanged`. As the name goes, whenever the X, Y, and Z values change, the accelerometer will fire this event and provide us with the updated values. So we need to attach an event handler to the `ReadingChanged` event. In our case, we will update the three text blocks in XAML with the updated values when the event is fired. The code snippet for it is shown as follows:

```
accelerometer.ReadingChanged.Add(fun args ->
  do this.Dispatcher.BeginInvoke(new Action(fun _ ->
    xTextBlock.Text <- "X: " + args.X.ToString("0.00")
    yTextBlock.Text <- "Y: " + args.Y.ToString("0.00")
    zTextBlock.Text <- "Z: " + args.Z.ToString("0.00")
  ) ) |> ignore
```

7. The `ReadingChanged` event exposes event arguments of type `AccelerometerReadingEventArgs`. This class provides us with the new values being read by the accelerometer. The reading event occurs on the accelerometer thread and we cannot update the UI from that thread. So notice the use of the `Dispatcher` class to paint the UI. The args has the X, Y, and Z properties, which can be read and will provide the latest readings. So we just update the text blocks with the new values.

8. In order to start recording the movement of the device we need to start the accelerometer. The accelerometer class has a `Start()` method. So we need to make a call to the `Start()` method:

```
accelerometer.Start()
```

9. Compile the solution to make it ready for deployment. To test the accelerometer, we can either deploy our app onto a physical device or use the emulator itself. The emulator comes with an accelerometer sensor simulator. The following steps will help you access the simulator:

   a. Run the demo and let the emulator come up.

   b. Move your cursor to the right of the emulator to display the emulator toolbar as shown in the following screenshot:

   c. Click on the last button at the bottom of the emulator toolbar.

   d. Click on the **Accelerometer** tab to view the accelerometer sensor simulator.

10. The following is an image of the simulator:

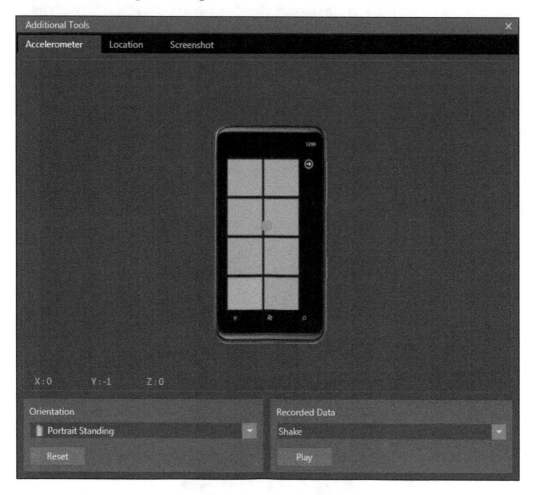

In the middle of the touch pad, drag the pink dot to simulate the movement of a device in a 3D plane.

11. The output of the app that we just created as the demo is shown as follows:

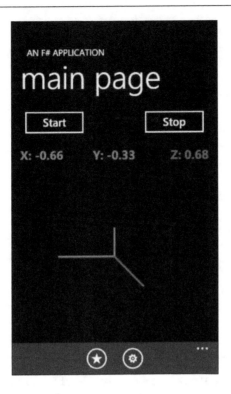

GPS (location services)

The Windows Phone operating system contains location service, which allows us to build location-aware applications. The service makes use of sensors, such as a GPS receiver, Wi-Fi, and cellular radio, which makes up the hardware on the device, to provide the location information. The location service can use GPS, Wi-Fi, or cellular radio to deduce the location while balancing the power-utilization performance. Location service exposes the phone's current longitude, latitude, altitude, speed of travel, and heading through the System.Devices.Location namespace.

Let's create a demo app to work with the location service. The following steps will help you create the demo:

1. Create a new project of the type **F# Windows Phone Application (Silverlight)**. Give it a name.

2. In the **App** project, right-click and add a reference to the System.Device assembly. The location service is provided in the System.Device.Location namespace.

3. Add a couple of text blocks in `MainPage.XAML`. We will try to read the latitude, longitude, speed, course, and altitude. The XAML snippet is given as follows:

```
<StackPanel Orientation="Vertical">
  <TextBlock Text="Lat:" />
  <TextBlock x:Name="txtLatitude" />
  <TextBlock Text="Long:" />
  <TextBlock x:Name="txtLongitude" />
  <TextBlock Text="Speed:" />
  <TextBlock x:Name="txtSpeed" />
  <TextBlock Text="Course:" />
  <TextBlock x:Name="txtCourse" />
  <TextBlock Text="Altitude:" />
  <TextBlock x:Name="txtAltitude" />
</StackPanel>
```

4. In the `AppLogic.fs` file, add a reference to the `System.Device.Location` namespace. This is required to access the location service:

```
open System.Device.Location
```

5. In `AppLogic.fs`, find the `MainType` class and add the following code:

```
let latitudeTextBlock : TextBlock = this?txtLatitude
let longitiudeTextBlock : TextBlock = this?txtLongitude
let speedTextBlock : TextBlock = this?txtSpeed
let courseTextBlock : TextBlock = this?txtCourse
let altitudeTextBlock : TextBlock = this?txtAltitude
let watcher : GeoCoordinateWatcher
 = newGeoCoordinateWatcher(GeoPositionAccuracy.High)
do watcher.MovementThreshold <- 10.0
do watcher.PositionChanged.Add(fun args ->
   latitudeTextBlock.Text <-  args.Position.Location.Latitude.
ToString("0.0000000000000")
    longitiudeTextBlock.Text <- args.Position.Location.Longitude.
ToString("0.0000000000000")
   speedTextBlock.Text <- args.Position.Location.Speed.
ToString("0.0") + "Meter/Sec"
   courseTextBlock.Text <- args.Position.Location.Course.
ToString("0.0")
    altitudeTextBlock.Text <- args.Position.Location.Altitude.
ToString("0.0")
   )
do watcher.Start()
```

What we are doing here is, create an instance of `GeoCoordinateWatcher`, set the movement threshold, that is, how much distance traveled by the user we need a notification for, and finally add an event handler to the `PositionChanged` event. The event handler receives an argument of type `GeoPositionChangedEventArgs`, which contains the `Position` property. This property will contain the new coordinates.

Now in order to test the location service, you can deploy the demo app on a physical device or use the Windows Phone emulator, which provides the location sensor simulator. Run the application and let the emulator come up. From the emulator toolbar, select the last button from the bottom to open the **Additional Tools** window. Click on the **Location** tab to view the location sensor simulator. A screenshot of the simulator is shown as follows:

The following steps will help you test the application with a live input:

- Turn on the Live input mode by toggling the **Live** button.
- In the search box, type a location and press *Enter*.
- Toggle the push-pin button to turn it on.
- Click on the map to add points or push pins on the map. Every time a pin is added, it will fire the `PositionChanged` event and the code should react to the event and show the new coordinates.

Summary

In this chapter, we saw how the hardware requirements enhance the user experience of the phone by making it mandatory to have certain sensors on he device. We looked at two of the most widely used sensors: accelerometer to detect the X, Y, and Z coordinates of the device, and GPS location service, which allow our application to be location-aware. Both of these sensors make it possible for the application developers to adopt them with ease and enhance the features of an application.

Index

Thank you for buying
Windows Phone 7.5 Application Development with F#

About Packt Publishing

Packt, pronounced 'packed', published its first book "Mastering phpMyAdmin for Effective MySQL Management" in April 2004 and subsequently continued to specialize in publishing highly focused books on specific technologies and solutions.

Our books and publications share the experiences of your fellow IT professionals in adapting and customizing today's systems, applications, and frameworks. Our solution based books give you the knowledge and power to customize the software and technologies you're using to get the job done. Packt books are more specific and less general than the IT books you have seen in the past. Our unique business model allows us to bring you more focused information, giving you more of what you need to know, and less of what you don't.

Packt is a modern, yet unique publishing company, which focuses on producing quality, cutting-edge books for communities of developers, administrators, and newbies alike. For more information, please visit our website: www.packtpub.com.

About Packt Enterprise

In 2010, Packt launched two new brands, Packt Enterprise and Packt Open Source, in order to continue its focus on specialization. This book is part of the Packt Enterprise brand, home to books published on enterprise software – software created by major vendors, including (but not limited to) IBM, Microsoft and Oracle, often for use in other corporations. Its titles will offer information relevant to a range of users of this software, including administrators, developers, architects, and end users.

Writing for Packt

We welcome all inquiries from people who are interested in authoring. Book proposals should be sent to author@packtpub.com. If your book idea is still at an early stage and you would like to discuss it first before writing a formal book proposal, contact us; one of our commissioning editors will get in touch with you.

We're not just looking for published authors; if you have strong technical skills but no writing experience, our experienced editors can help you develop a writing career, or simply get some additional reward for your expertise.

Windows Phone 7.5: Building Location-aware Applications

ISBN: 978-1-849687-24-9   Paperback: 148 pages

Build your fi rst Windows Phone application with Location and Maps

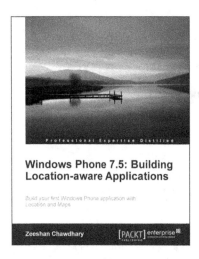

1. Understand Location Based Services.

2. Work with Windows Phone Location Service.

3. Understand how Maps work.

4. Create a simple Map application and learn to use Geocoding, Pushpins.

Windows Phone 7.5 Data Cookbook

ISBN: 978-1-849691-22-2   Paperback: 224 pages

Over 30 recipes for storing, managing, and manipulating data in Windows Phone 7.5 Mango applications

1. Simple data binding recipes to advanced recipes for building scalable applications

2. Techniques for managing application data in Windows Phone mango apps

3. On-device data storage, cloud storage and API interaction.

Please check **www.PacktPub.com** for information on our titles

Windows Phone 7 XNA Cookbook

ISBN: 978-1-849691-20-8   Paperback: 450 pages

Over 70 recipes for making your own Windows
Phone 7 game

1. Complete focus on the best Windows Phone 7
   game development techniques using XNA 4.0

2. Easy to follow cookbook allowing you to dive in
   wherever you want.

3. Convert ideas into action using practical recipes

Windows Phone 7 Silverlight Cookbook

ISBN: 978-1-849691-16-1   Paperback: 304 pages

All the recipes you need to start creating apps and
making money

1. Build sophisticated Windows Phone apps with
   clean, optimized code.

2. Perform easy to follow recipes to create
   practical apps.

3. Master the entire workflow from designing
   your app to publishing it.

Please check **www.PacktPub.com** for information on our titles

www.ingramcontent.com/pod-product-compliance
Lightning Source LLC
LaVergne TN
LVHW080059070326
832902LV00014B/2317